Deceiving (Dis)Appearances

Analyzing Current Developments in European and North American Border Regions

P.I.E. Peter Lang

Bruxelles · Bern · Berlin · Frankfurt am Main · New York · Oxford · Wien

Harlan KOFF (ed.)

Deceiving (Dis)Appearances

Analyzing Current Developments in European and North American Border Regions

"Regionalism & Federalism"
No. 12

This volume derives from a workshop, organized by the editor, with the support of the Robert Schuman Centre for Advanced Studies at the European University Institute.

The editor wishes to thank the Robert Schuman Centre for Advanced Studies at the European University Institute and the University of Luxembourg for their financial support of this project.

© P.I.E. PETER LANG s.a.
Éditions scientifiques internationales
Brussels, 2007
1 avenue Maurice, B-1050 Brussels, Belgium
info@peterlang.com; www.peterlang.com

ISSN 1379-4507
ISBN 978-90-5201-369-5
D/2007/5678/56
Printed in Germany

Bibliographic information published by "Die Deutsche Bibliothek"

"Die Deutsche Bibliothek" lists this publication in the "Deutsche Nationalbibliografie"; detailed bibliographic data is available in the Internet at <http://dnb.ddb.de>.

CIP available from the British Library, GB
and the Library of Congress, USA.

Contents

Preface

The present book is the result of a workshop supported by the Robert Schuman Centre of the European University Institute on the empirical analysis of so-called "territorial challenges to the nation-state caused by globalization". Specifically, this phrase refers to sub-national and transnational institutions, markets, identities, etc. that scholars and politicians claim have created challenges to the legitimacy of the nation-state. Moreover, much attention has been paid to activities of supranational bodies, such as the European Union and the Council of Europe, aimed at establishing and reinforcing regional and cross-border institutions and identities.

While theoretical debates surrounding the present ability of nation-states to protect their borders and guarantee their sovereignty are well developed, there has been a lack of comparative empirical studies that demonstrate real transformations in border regions and their significance to sovereignty issues. In addition, scholars have often examined the activities of supranational organizations aimed at promoting sub-national and transnational development in these areas, but little policy evaluation has been carried out to determine the actual impact of these policies. The participants in this workshop presented empirical research from Europe and North America that attempts to fill these gaps in contemporary debates on regionalism, both sub-national and transnational.

In addition to the authors who contributed chapters to this volume, I would like to thank numerous people who helped make this project a success. First, I would like to thank Helen Wallace, the former director of the Robert Schuman Centre. She made funding available for Jean Monnet Fellows to organize workshops at the Centre. This initiative was appreciated by many of us and it helped further embed the work of visiting fellows into the activities of the European University Institute. I would also like to thank Dr. Wallace for her constructively critical comments on my own research which have significantly improved the quality of my current project and for her general support of my comparative research agenda.

Second, I would like to thank Michael Keating for participating in the above-mentioned workshop and for generally supporting the project.

This publication is the fruit of his collaboration. I am very appreciative for his backing and for his patience with the manuscript.

Third, I would like to thank many people who made significant contributions to this project which are not overtly highlighted in this volume. Leonardo Morlino, Alessandro Silj, Claudia Carolina Zamorano-Villareal, Adrian Farell, Monica Sassatelli and Ulrich Sedelmeier all made scientific contributions to our discussions in the workshop which stimulated thought and debate. Hervé Dupuy came to the workshop as the keynote speaker from the European Commission and left as a valued friend. Eros Cruccolini, the President of Florence's City Council and Moreno Biagioni, of the *Associazione Nazionale Comuni Italiani*, received the workshop participants at the Palazzo Vecchio and organized a roundtable on border politics with members of local non-governmental organizations. I thank both of these gentlemen for their specific contributions to this project and their warm hospitality, for their general dedication to world improvement, and for being a friend to all. Laura Burgassi and Mei Lan Goei provided invaluable adminsitrative support for which I am greatly appreciative.

I would also like to thank my family who has been pushing me across borders all of my life and supporting me throughout the journey, and my wife Carmen, whose love knows no boundaries. She has taught me through her words and actions that borders are not only to be surpassed, but also to be redefined on one's own terms. Her intellectual contribution to this volume goes well beyond the chapter on which her name appears. Finally, I would like to thank our daughter Zenyram… for being Zenyram.

Harlan Koff

INTRODUCTION

Power, Politics and Deceiving (Dis)Appearances

Harlan KOFF

I. Introduction: The Contemporary Re-definition of Borders

Borders have been significantly re-defined since the early 1990s. Of course, substantial political and economic changes have occurred during this period due to the fall of authoritarian regimes in Central and Eastern Europe as well as Latin America, the deepening and enlargement of the European Union (EU), the development of regional integration in many other parts of the world (such as the North American Free Trade Agreement (NAFTA), MERCOSUR, the Andean Community and the Association of Southeast Asian Nations (ASEAN)), and the increased exchange speed and volume of international trade and information. The impact of such shifts in global geopolitics and economic markets has also led to the re-conceptualization of national borders. In order to respond to so-called "threats" to the sovereign control states hold over their boundaries, scholars have shifted their analysis away from the narrow idea of "borders", and they have moved their attention towards the wider view of "borderlands", "border regions", "border zones", and "transnational communities", thus, leading to the conceptual re-definition of border politics.

These recent approaches have identified border areas as socially constructed territories that demonstrate many of the characteristics of independent polities. For this reason, recent scholarship has examined the creation of cross-border political institutions, border identities, the expansion of border markets, and the development of cross-border political mobilization. Border communities seem to have come to life, creating a degree of autonomy and separation from central state actors.

While the rich literature (see review below) in border studies identifies many important changes in political and economic systems in these areas, it does not necessarily identify the mechanisms that create these changes. Social scientists are often confronted with the dilemma of linking the explanation of individual cases to the construction of general theory. In the emerging field of border studies, this has become one of the most recognizable challenges. Many scholars, such as James Anderson, Blatter, Brunet-Jailly, Newman, and Perkmann and Sum, have recently attempted to formulate "border theory" which tries to both explain the empirical variance in integration levels found in different border areas, and include the diverse theoretical lenses utilized by geographers, urban planners, political scientists, anthropologists, sociologists and economists active in this field of study. Further complicating the pursuit of this goal is the fact that most scholarship of the European Union has developed through the importation of traditional analysis of domestic political systems (i.e. Hix), whereas the study of regional integration in other parts of the world has focused on theories of international relations with particular focus on international political economy (i.e. Mattli). Because border politics are so closely tied to regional integration processes, this has led to a divergence in analytical approaches and a lack of inter-continental empirical scholarship (see Blatter, Scott).

Despite the well-developed theoretical debates surrounding the evolution of border communities, discussions on this topic are characterized by two main weaknesses. First, it often seems that authors are forwarding normative judgements concerning the need for heightened border integration rather than explaining the processes that cause it. Second, the complexity of border politics has made theory-building extremely difficult. Some basic questions that challenge empirical analysis are: "Why has integration occurred in some border regions while other borders are being reinforced? Why has integration failed in some cases where opportunity structures are positive, where it has succeeded in others saddled with more limited constraints? Can border integration be achieved through a 'top-down' approach or is it necessarily a 'bottom-up' process?"

The purpose of this volume is to address these weaknesses in debates concerning contemporary border politics. Rather than attempting to create a global interdisciplinary theory that cannot explain the disparity of empirical cases, the book attempts to isolate the mechanisms that link international forces, such as regional integration and economic globalization to political outcomes at the sub-national level through comparative analysis based on a selection of most different cases.

First, this book presents empirical research conducted in Europe and North America. Because it varies the continental context, it examines political outcomes in border communities under different regional integration models (EU and NAFTA) which have diverse characters and

objectives. Second, the contributions to this book analyze border politics within different spheres, including ethnic identity and mobilization
(Bray and De Frantz), cross-border political and economic cooperation
(Obydenkova and Scott), environmental politics (Maganda) and border
security (Koff and Sabet). By comparing different policy arenas, we can
isolate relevant input variables in the political systems examined. Third,
the book includes cases that represent both "disappearing" (Bray, Scott
and De Frantz) state borders that benefit from transnational initiatives at
the supranational level, "reinforced" borders (Obydenkova, Koff, Sabet)
where separation is increasingly pronounced, as well as one paper that
compares both national (Mexico-United States) and sub-national
(Lerma-Chapala Water Basin) borders. Thus, this book varies the types
of political borders examined. According to both political strategies and
academic analyses, border integration should be more easily achieved
across internal borders than external ones. Moreover, integration should
be more evident in issue arenas related to economic development and
civil society than it should be in other fields, such as security politics
and environmental cooperation. The papers presented here show that
this is not necessarily the case. By doing so, they question many of the
presumptions that dominate the current literature on border politics.
These premises are described in the following section.

II. The Current Boundaries of Border Research

Research on borders has blossomed since the beginning of the
1990s. In his *Progress in Human Geography* lecture at the April 2005
annual meetings of the Association of American Geographers, David
Newman noted that numerous internationally renowned journals have
dedicated special issues to the study of borders; two major associations
have been born (the Association of Borderland Studies and the International Boundaries Research Unit); numerous academic networks, such
as BRIT (Border Regions in Transition) have appeared that focus on
this issue; and one academic publisher has a book series dedicated to the
study of borders.[1] Moreover, the *Journal of Borderland Studies* has
promulgated important international debates in border research.

Traditionally, border studies focused on the historical relevance of
borders to domestic politics and international relations more than they
examined developments in border communities themselves. Borders
were often analyzed as boundaries that not only provided lines of de-
marcation between political units, but also cultural or economic distinc-
tions between so-called "civilization" and backwardness. Historical

---

[1] Newman, D. 2006. "The lines that continue to separate us: borders in our 'borderless'
world". *Progress in Human Geography* 30 2, p. 2-3.

approaches have noted how borders in the Roman Empire or Medieval Europe created a hierarchy of spaces which focused on the control of specific territories (Brunet-Jailly). Borders were often perceived as points of conflict and analysis of these areas often concentrated on their relevance to global geopolitical issues as they were simply framed as the places "where politics occurred". For example, in his study of political conflict in Latin America, Michel Foucher created a typology of borders that included a) "immediate borders" that demarcated the division between conjoined states, b) "imperial borders" that separated the hemisphere's superpower, the United States, from the rest of Latin America, and c) "areas of fracture" that differentiated between the developed and the developing world.[2] Despite his empirical focus on specific geographic zones, Foucher's theoretical analysis, like that presented by many of his contemporaries, remained fixed on the international arena.

Historians have also noted that the differences between European and North American notions of borders have their roots in the distinct demographic and economic development of these continents. For example, studies of border vocabulary, (i.e. Malcolm Anderson) have described how the French term *frontière* refers to the end of a political territory, which raises potential for conflict with those on the other side, whereas the American version of *frontier* has connotations linked to shifting patterns of human settlement, the expansion of economic opportunity, and the spread of political values. This latter interpretation, in fact, moves the focus of border analyses from the significance of border regions in international politics to the border communities themselves. The recent boom in borderland studies has similarly transferred attention from the international arena to the national and local levels, creating a sphere of micro-scaled spatial study.

This trend is related to three important shifts in the study of international relations that significantly affect border areas. All three examine so-called "challenges to the sovereignty of nation-states". These new scholarly paradigms include: a) "multi-level governance", b) the "borderless world", and c) "new regionalism". Each will be discussed in the following sub-sections.

Multi-level Governance

Whereas traditional analysis of regional integration, especially the European Union, focused on the political tension between supranational organizations and their member states in political debates regarding

---

[2] Foucher, M. 1986. *L'Invention des Frontières*. Paris: La Documentation française, p. 233.

authority in policy arenas, recent studies have begun analyzing institutional relationships through the lens of "multi-level governance". This framework, originally proposed by Gary Marks and Liesbet Hooghe, argues that successful policy implementation, especially in the arenas of regional and social development is dependant on the activities of local government. The authors describe two types of multi-level governance. Type 1 analyzes the standing relationship between agencies of general purpose jurisdiction. Conversely, Type 2 multi-level governance is understood as the interaction of public and private local, national, and supranational actors in specific policy processes, such as security, economic development, environmental protection, etc.

More recent empirical studies of EU regional policies have focused on both types of multi-level governance. Institutional studies, such as those conducted by Perkmann, Aykaç, Benington and Harvey, Balme, etc., have examined Type 1 governance through discussions of EU programs, such as INTERREG initiatives, in cross-border regions. Perkmann has noted that over seventy border communities (both cities and regions) have entered into some sort of formal organizational agreement with counterparts on the other side of national divides.[3] He, like others (i.e. James Anderson), has also raised doubts about the effectiveness of these programs. While many observers argue that cross-border cooperation provides both an avenue for political emancipation from central authorities and more efficient policy implementation, Perkmann correctly notes that no comparative empirical evidence exists to prove these claims. In fact, some European studies (see De Frantz and Bray in this volume) demonstrate that cross-border policy-making does not always create expected outcomes.

Similarly, one does not always observe expected outcomes when analyzing policy-making in North American border regions. Of course, because NAFTA does not include regional or social policies to the extent of the European Union there are no formal strategies that correspond to European initiatives for border development, such as INTERREG, Corridor 8, etc. Moreover, the borders that divide Canada, the United States and Mexico have been reinforced since September 11, 2001. For this reason, policy studies concerning North American borders have retained, at least partially, the focus on borders as spaces where international politics occur (see Papademetrious, Andreas, Cornelius). This is especially true in terms of security politics, where US policies have aimed to stop external threats by controlling national

---

[3]   Perkmann, M. 2003. "Cross-border Regions in Europe. Significance and Drivers of Regional Cross-border Co-operation". *European Urban and Regional Studies* 10 (2), p. 155.

borders. As a result, the US border with Canada, which used to be considered a "soft" border that could be easily penetrated, has become as fortified as the US-Mexico divide, where immigration patrols have been active for years. In fact, Immigration and Customs Enforcement has received exponential increases in both its budget and personnel to become the largest agency in the US federal government (Cornelius). This reinforcement of NAFTA's internal borders has impeded the development of Type 1 multi-level governance because the structures necessary to cultivate it do not exist.

Despite US efforts to control both of its borders, much binational cooperation does occur at the local level in North America and many studies have been conducted of this activity. In fact, cooperation between the public and private spheres is fundamental to the understanding of North American border politics due to prevailing notions of governance on the continent. For years, the American (in the continental sense) frontier was ignored by national policy-makers so local communities created informal working relationships to address common socio-economic problems. In his study, *US-Mexico Borderlands*, Oscar Martinez examines the historical development of these areas and argues that competition and hostility transformed into peaceful coexistence over time, which has since led to cooperation and management due to the interdependence of these communities.[4] This cooperation takes place in specific spheres and usually spills-over into other policy arenas. Once informal local collaboration occurs, it is often ratified later through binational agreements (Type II multi-level governance).

Many empirical studies of specific border communities have reinforced this argument. In fact, the dearth of national or supernational policies in border planning has provided sub-national governments with the freedom to cooperate on cross-border strategies for economic and social development. Scholars, such as Alper, Clarkes, Erie, Hanson, Herzog, etc. have demonstrated how local communities at both the US-Mexico and US-Canada borders have benefited from weak control at higher levels of government to collaborate on environmental protection strategies, economic development policies, etc. Carmen Maganda furthered this argument, in her recent analysis of border water policies where she demonstrated that the implementation of NAFTA, and its side agreements concerning water-sharing actually provided a legal/institutional foundation for political conflict, whereas local authorities along the US-Mexico border had previously worked together on a

---

[4] Martinez, O., ed. 1996. *US-Mexico Borderlands: Historical and Contemporary Perspectives*. Wilmington DE: Jaguar Books, p. xiii-xix.

daily basis to construct common-sense solutions to water scarcity problems that affected them all.

For many observers, such "bottom-up" cross-border cooperation is a sign that nation-states can no longer control flows across their national borders as local actors follow their own political and economic interests. This has contributed to a related argument that increased competition to central state authorities has created a "borderless world".

Globalization and the "Borderless World"

Many authors have contended that we will soon live in a world where borders will have minimal importance, should they exist at all. Of course, this position extends well beyond the planning arena discussed above. In fact, there are few examples of borders that have created unified cross-border institutions. Most "border integration" is viewed in terms of cross-border cooperation and economic linkages. Rather than a single integrated body, border communities often create increased interaction between autonomous polities.

Thus, the foremost proponents of "borderless" politics are not schol-ars of border communities themselves, but social theorists discussing international geopolitics. Prominent supporters of this argument include Ohmae, Castells, Sassen, Jacobson, Joppke, Bauböck, etc. These schol-ars argue that socio-economic activities are no longer constrained by state borders and transnational networks of citizenship, information, intellectual exchange, social mobilization, etc. have created competition to the nation-state. Moreover, political economists, such as Rogowski, Frieden, etc. contend that economic competition has been transformed in the post-Fordist era. Whereas previous cleavages divided states or economic classes, these authors argue that contemporary divisions are based on geographic characteristics (i.e. rural *versus* urban) or economic sectors (i.e. traditional industry *versus* high technology), thus decentral-izing economic relationships. They conclude that the emergence of this new organization of the global economic system has further eroded the ability of the nation-state to control the activities of non-state actors.

The more that borders themselves have "withered away", the more significant border regions have become in domestic and international politics. David Newman correctly argues that: "For some, the notion of a 'borderless' and 'deterritorialized' world has become a buzzword for globalization".[5] It must be noted, however, that the emergence of cross-border communities are only one example of the so-called de-territorialization of the nation-state. Scholars of high-skilled migration,

---

[5]   Newman, p. 1.

such as Favell and Smith, have noted that increased migration among socio-economic elites has created a cosmopolitan system of citizenship and identity that opposes ethno-nationalism. Thus, according to these theories, de-territorialization is challenging nation-states from both above (i.e. economic systems, global information networks, etc.) and below (i.e. cross-border social mobilization). Of course, these structural changes in socio-economic and political systems have affected identity politics, especially in border areas. This is the focus of a different paradigm that has significantly influenced contemporary border studies.

New Regionalism

According to recent scholarship on political identities, the de-territorialization of the nation-state has led to the emergence of new ideologies and movements based on ethnicity, language, place of belonging, etc. In some cases, such as Northern Italy where the Lega Nord political party has invented the notion of an autonomous 'Padania', (see Diamante) these new movements are politically constructed identities that serve non-ethnic political goals. In most other geographic areas, however, emerging ethnic regionalism has focused on cultural claims for recognition or autonomy. Leading scholars in this field, such as Michael Keating, have demonstrated that the nature of local political organization and culture has significantly affected how border politics function. Studies of nationalist movements (such as the Catalan movement), minority groups (such as Gypsies in Eastern Europe), and state-less nations (such as the Québecois or Basques) have examined the connections between these groups and the global political arena. In his study of pluri-nation states, Keating has demonstrated that these movements are not necessarily integrated into the institutional architecture of their respective countries. Thus, territorial boundaries are not fundamental for the construction of identity, which decreases the demarcating influence of borders within ethnic debates. For example, some groups, such as the Scots or Corsicans, correspond to an existing territory and they benefit from special representative structures. Others, such as Basques, the Flemish, Gypsies, etc. adhere to de-territorialized identities that cross state borders.

Support for these movements has developed at the supranational level and, in some instances this has directly challenged central state authorities. For example, through support from the Council of Europe (COE), non-governmental organizations (NGOs), such as the European Roma Rights Center (ERRC), have successfully created a transnational Roma identity during the last twenty years, even though Roma represent one (albeit the largest) ethnic identity amongst European Gypsies, along with Sinti, Manouche, etc. As it gained strength, this political movement solidified working relationships with various European institutions. The ERRC has successfully brought cases to the European Court of Human Rights (ECHR) in order to combat state-sponsored anti-Gypsy discrimination throughout the continent. These judgements in their favor have further legitimized the political aspect of this de-territorialized identity.

In North America, influential supranational structures such as the COE and the ECHR do not exist and this has created difficulties for the development of transnational political movements. While some observers, such as Brooks and Fox, Williams, etc. have studied the creation of transnational social movements, and others, such as Bonilla *et al.* have examined the historical development of cross-border cultures through the presentation of border narratives, most political participation along the US-Mexico or US-Canada borders has been characterized by blocked exchanges. For example, NAFTA's stimulation of the maquiladora[6] border industry has created numerous movements in both the US and Mexico, that have called for improved working conditions, higher wages, and social justice. Nonetheless, most of these movements do not carry out binationally cooperative activities.

This separation has been created by two important factors. First, the US reinforcement of the border has limited opportunity structures for cross-border mobilization (see Andreas). Second, Mexicans and Mexican-Americans do not always congregate in a unified political movement. Because many Mexican-Americans or Chicanos, no longer speak Spanish, they are not always accepted by Mexicans. Moreover, these movements are divided over the immigration issue. Whereas Mexican political entrepreneurs focus their efforts on migrants' rights, many Mexican-Americans have supported immigration controls in order to separate themselves politically from the clandestine waves of migrants arriving from Mexico. In fact, the spring 2006 protests against the construction of a 3,000 kilometre wall along the entire US-Mexico border are the first evidence of widespread binational political protest (Koff). This demonstrates that borders have a more divisive role in North America than they do in Europe, not only because they block the passage of people, goods, etc. or delineate political territories. Instead, their most important effects seem to relate to the limits that they place on opportunity structures in border regions. For this reason, one would expect to find significantly deeper levels of cross-border cooperation in

---

[6] *Maquiladoras* are bonded assembly plants found in Mexican border cities that are permitted to import goods without payment of import duties. These goods, especially electronics, are further processed or manufactured and exported. When the goods enter the US, tariff is levied only on the value added outside the US.

Europe than in North America. This is, in fact, the subject of this volume and the focus of the following comparative introduction.

III. Deceiving (Dis)Appearances in Border Regions

As the literature review presented above illustrates, the field of border studies has moved towards paradigms that, to varying degrees, examine the distinctiveness of border regions and their increased autonomy from central actors. Of course, empirical evidence does not seem to support exaggerated neo-liberal conclusions that point to the erosion of borders and thus, nation-state sovereignty. However, it could support the claim that borders have become increasingly permeable through the selective lowering of various barriers. Most of the studies that point to this conclusion (see above) view this outcome as the result of actions taken by local leaders to liberate themselves from central state control. This premise, on which many border studies are based, must be questioned. Who are these "local leaders" who "logically" seek increased autonomy? Why is it necessarily in the interests of border communities to distance themselves politically from central state authorities? Why is cross-border collaboration assumed to occur, especially within the EU, just because supranational initiatives attempt to foster it? These questions are the focus of this volume, which opens the black box of decision-making processes in border regions in Europe and North America.

The Political Economy of Border Regions

When discussing globalization or transnationalism, most studies (i.e. Ohmae) implicitly suggest or explicitly contend that there has been an integration of the political and economic spheres. A borderless world necessarily suggests that political actors serve the creation of cross-border markets by lowering formal and informal barriers to free movement. Contemporary scholars of borders, such as James Anderson, argue that borders are no longer meant to control transnational flows as much as they are filters for these economic and demographic currents.

Of course, the logic at the base of this system varies both geographically and sectorally as certain arenas, especially those related to economic development, are more prone to transnationalism than others, such as education or welfare politics. For this reason, James Anderson correctly suggests that most border theory is, in fact, characterized less by the integration of politics and economics and more by the separation of these spheres. He writes,

> Rosenberg (1995) following Wood (1994), makes the insightful argument that 'the separation of the economic and the political in capitalism' became a structural necessity for territorially delimited political sovereignty (untram-

melled by 'economics') and for the globalisation of economic production (unconfined by territorial 'politics' and state borders).[7]

This argument conforms to most general approaches in the field of political economy. For example, Caporaso and Levine argue that political economy does not refer to the integration of politics and economics, but it encompasses the relationship between these two distinct arenas. Anderson similarly concludes that: "'politics' do not simply 'stop' at borders nor do 'economics' simply 'speed through'."[8]

Anderson's approach is valid but it lacks one important element: the identification of the mechanism that explains the primacy of economic decision-making in border areas. He argues that "the general historical tendency... has been for transnationalism to depend less and less on formal empires and increasingly to involve looser, more informal modes of imperialism or non-hierarchical cross-border relations".[9] By introducing such concepts he implicitly refers to the notion of power and how it has evolved over time. While he correctly identifies the need to better understand the relationship between politics and economics in border areas, he presents a vision of 'politics' that is statist and limited.

In order to better understand the relationship between politics and economics, this volume presents studies in different arenas that examine the decision-making processes found in border communities. Rather than utilizing a policy-based approach, the volume analyzes border politics through the input side of political systems. It employs a notion of political economy that is not based on 'economics as a field or market arena' but rather it comprehends 'economics as a method' that employs rational decision-making based on cost-benefit analyzes.

This volume argues that border integration is not simply a logical outcome of globalization or multi-level governance. Instead, it occurs because political leaders, both governmental and non-governmental, are rational actors that compete within political and economic systems for various resources. This point is especially important because much discussion focuses on local institutions. These institutions, however, are not simply structures that regulate the behavior of political actors (see North). Instead, political institutions are characterized by 'interests' because they are run by decision-makers that follow their own individual incentives (see Levi).

---

[7] Anderson, J. 2001. "Theorizing State Borders: 'Politics/Economics' and Democracy in Capitalism". *CIBR Working Papers in Border Studies* CIBR/WP01-1, p. 5.

[8] *Ibid.*, p. 8.

[9] *Ibid.*, p. 5.

Thus, border communities are not unlike other polities where political decisions are based on the short-term interests of political entrepreneurs. Anderson, himself, concludes:

> Even where borders as 'gateways' between separate polities are seen as a source of positive economic opportunities their consequences at a deeper level are arguably negative. Borders tend to generate short-term opportunistic and more questionable 'argitrage' activities, ranging from trading on tax and price differences to smuggling and associated forms of crime, including the smuggling trade in 'illegal immigrants'.[10]

Hence, the argument presented in this volume is that border integration is the result of political bargaining within a dynamic political landscape characterized by shifting opportunity structures.

This point addresses the research questions posed in the introduction. It explains variance in levels of comparative border integration. It also discusses the process of institutional border integration, the cross-border expansion of civil society, and the relationship between political institutions and economic markets. Border integration in these diverse arenas can be understood as a product of political entrepreneurship. When entrepreneurs, including elected officials, NGO representatives, businesses, economic organizations, unions, etc. find it in their interests to create cross-border communities, then integration occurs, even when external conditions are not favourable (i.e. migrant rights movements in North America). Conversely, when favourable conditions exist due to the presence of supranational programs, such as those enacted by the EU, border integration does not necessarily come about because local actors do not rationally find these strategies to be coherent with their individual interests. This point is further developed in the next section.

Political Competition in Border Communities

The papers presented in this volume discuss different aspects of border politics and the connections between them may not be immediately evident. However, each of the papers are linked through their discussion of decision-making processes, their focus on competition for the attainment of political or economic resources, and their interest in the nature and uses of power in border areas.

Institutional Arrangements and Border Integration

Part one of the book examines institutional development and cross-border regionalization projects in three different contexts. First, James Scott discusses the spatial creation of cross-border regions through the

---

[10]   *Ibid.*, p. 9.

analysis of local transformation contexts at the German-Polish and Austrian-Hungarian borders, and how they interact with EU and national policies. Second, Anastassia Obydenkova examines EU-Russia relations through the comparative analysis of EU border policies and their impact on non-EU (Russian) border communities. Third, Carmen Maganda studies the relationship between regions, borders and states in North American water politics by comparing the allocation of water resources in transboundary Basins (the Lerma-Chapala Basin in Central Mexico and the Colorado River Basin that crosses the US-Mexico border). These chapters represent the analysis of institutional arrangements at two internal EU borders, two external EU borders, and the comparison of national and sub-national borders in North America.

The three cases presented obviously address border cooperation in different geographical and institutional contexts. More importantly, however, is the fact that they discuss cases where local officials face disparate incentives for border integration. Despite these differences, the cases are linked by the fact that political actors follow similar patterns of behavior despite the presence of these disparate institutional frameworks. For example, the two cases presented in James Scott's chapter are internal EU borders. Both of the cases are integrated into EU regionalization strategies; they share common socio-economic conditions; their core-periphery relationships are similar; and both are characterized by large income and GDP difference. "Why then do officials in West Pannonia (Austria-Hungarian border) engage in more cross-border initiatives than their German-Polish Euro-Region counterparts?" Scott presents strong evidence that "Cross-border regionalism does not exist for its own sake. It requires a rationale, whether economic, environmental, and/or political". Scott's chapter clearly demonstrates that informal incentives are fundamental elements for collaboration that he claims is based on "a learning process". Institutional arrangements that are imposed from above are perceived as contrived by political elites and community members and their presence does not guarantee cross-border integration.

Similarly, Anastassia Obydenkova examines the development of cross-border cooperation between the EU and two Russian regions and its impact on sub-national regime transition. She argues that EU policies not only affect neighbor countries, but they also impact the regions within those states. More specifically, Obydenkova examines the political strategies of regional actors in St. Petersburg and Leningrad oblast. Like Scott's research design, Obydenkova has chosen cases with significant structural similarities. Both border regions democratized during the same time period, they received similar amounts of autonomy in the asymmetrical federal system of government in Russia and both regions have long external borders with similar geographic characteristics. Despite these common qualities, the leaders of the respective regions have followed very different patterns of development and they have demonstrated very different levels of receptiveness to collaboration with the EU. Obydenkova explains this discrepancy through her discussion of the different economic bases of the regions. St. Petersburg followed an international path toward development with open markets, which led to the informal introduction of democratic values through EU foreign aid at the sub-national level. Conversely, leaders in Leningrad isolated themselves economically and turned toward Russian central authorities for economic and political support. Like Scott's chapter, this comparative study suggests that informal political competition and the rational pursuit of short-term objectives by border community leaders explain significant variance in levels of border integration. Moreover, both studies demonstrate that the creation of cross border civil society is possible when separate cross-border polities share political interests and visions. When this is not the case, cross-border civil society cannot be invented artificially from above and the impact of supranational border initiatives on democratic practices within border cases is extremely limited.

This argument is carried even further in Carmen Maganda's study of border water politics in North American basins. Maganda argues that new cross-border distribution structures need to be developed in order to institutionalize democratic principles in cross-border water basins. Maganda compares the Lerma-Chapala Basin within Mexico with the Colorado Basin that spans the US-Mexico border. Her analysis shows that the basin councils and administrative commissions charged with overseeing the distribution of water resources in both regions are characterized by the same structural flaws even though they operate at different levels of government. In both cases, the actors within the regions attempt to maximize their individual shares of water rather than cooperate on long-term, sustainable strategies that would guarantee water resources for all, in both the present and the future. In fact, Maganda argues that these governing structures were created with the idea of introducing representative bodies that would ensure the democratic distribution of water. Instead, they destabilized political systems, where formal authority regarding the distribution of water became an object of discussion. Also, civil society, where informal solutions to common water scarcity problems had previously been bargained on a daily basis by local officials, became less influential. Like Scott and Obydenkova, Maganda contends that the introduction of border integration structures at both the national and sub-national level, did not necessarily cultivate the creation of cross-border civil society. In fact, she argues that the opposite is true. Cross-border initiatives actually introduced further incentives for competition into cross-border water systems and therefore, they had a negative impact on both decision-making and civil society because they created heightened levels of tension and more self-interested behaviour. This topic is further addressed in part two of this volume.

Border Politics, Ethnic Competition and Civil Society

Questions regarding cross-border civil society and grassroots mobilization are addressed more directly in part two of this volume. As the literature review presented above indicates, one of the major themes of contemporary border studies is the creation of de-territorialized identities. Politically, two very different processes are taking place in Europe and North America. In the former, supranational organizations, such as the EU and the COE, are investing significant sums of money to stimulate cross-border mobilization, especially in areas where transnational minorities are concentrated. Conversely, in North America cross-border movements have emerged as a reaction to the negative social consequences of NAFTA in border regions (see previous discussion of *maquiladoras*) and US border security policies. In both instances, the comparative advantages of transnational social mobilization are presented as facts more than interpretations.

The two chapters included in this part of the book contest this approach. In Chapter five, Zoe Bray presents an analysis of the impact of European policies on Basque territorial politics. Some members of this ethnic group, of course, have forwarded one of the most high profile nationalist challenges to the nation-state in Europe, due to the political violence perpetrated by the radical group, ETA. For this reason, amongst others, Basques have received significant attention on the supranational agenda in Europe as both the European Union and the Council of Europe have fostered cross-border initiatives in the Basque region.

Bray's anthropological study of politics in the Basque territory openly questions the notion that a single Basque nationalist identity exists. Despite the presence of European policies that address "Basque issues" and academic studies that examine "Basque culture", Bray argues that even though cross-border relations have intensified in recent years, practical cooperation between Basques in France and Spain has not yet fully developed. Bray challenges the notion of a cross-border Basque political lobby or interest group. Through extensive fieldwork with political leaders on both sides of the border, Bray acknowledges the fact that these actors follow political agendas that do not cross state borders and their actions and objectives are dictated by different political and institutional contexts in Spain and France. Thus, European programs have done little to join Basque communities into a single polity with unified political objectives. The leaders of the separate Basque communities do not seem to recognize the comparative advantage of cross-border collaboration for the attainment of their short-term goals.

The following chapter, by Monika De Frantz, also examines national minority mobilization, but from a political science rather than anthropological perspective. Like chapter one (Scott), this contribution analyzes cross-border cooperation at Austria's Eastern border. Her chapter examines how the economic integration efforts of governmental elites in the border region of Burgenland are actually embedded within a common border identity. De Frantz examines how the changing border context of EU enlargement has revived national minority mobilization and whether this might facilitate cultural integration across borders or promote regional development.

Focusing on the cultural aspects of European cross-border cooperation, De Frantz shows how the border situation can turn from a historical weight and a socio-economic problem into symbolical capital for regional economic development policy. Nonetheless, she notes that beyond the symbolic acknowledgement of transnational border identities, the cross-cultural knowledge and the long-established contacts between minority organizations are actually under-used in official cross-border cooperation. The intercultural integration strategy promoted by governmental elites is actually limited by their own narrow territorial self-interest and the general mistrust between the populations on both sides of the border. Thus, the changing border context has had varied cultural effects. While political and economic improvements for minorities created by regional integration have favored the Austrian side of the border, these transformations have actually weakened the influence of ethnic minority associations in the region and revived pan-nationalist ideas promoting ethnic separatism beyond the state borders in the urban areas around Vienna. Ethnic cross-border activism, however, remains uncoordinated and finds little structural support from the EU or Austria as state borders still seem to constitute territorial and institutional barriers for ethnic mobilization and regional cooperation.

Border Reinforcement versus *Cross-border Cooperation: Contemporary Security Agendas*

The final part of this book, examines border security politics in Europe and North America. Numerous studies have already shown that this is the policy arena where border reinforcement is most prominent. As stated above, US border policies have effectively closed the national divides with both Canada and Mexico in an attempt to control external security threats. The European Union has followed similar border strategies as the EU has reinforced its commitment to the supranational control of the Union's external frontiers since the 2002 Seville meeting of the European Council. However, unlike the US, the EU has attempted to reinforce these control policies with cross-border cooperation strategies at the sub-national level.

In Chapter seven, Harlan Koff compares cross border cooperation in Europe and North America through a discussion of Italian-Albanian and US-Mexico cross-border development policies. His comparative study contends that cross-border integration is not necessarily a product of regional integration strategies. While these initiatives certainly define opportunity structures for local leaders, they are not sufficient to bring about integration themselves. In the case of San-Diego, CA, USA and Tijuana, Mexico, the disparate natures of the cities' economies, combined with the historical political aversion to cooperate found in each case has offset any integration advances made in civil society. Conversely, Bari, Italy and Durres, Albania have embraced integration because the cities share similar economic markets, they have historical ties (like San Diego-Tijuana), and most importantly, each city has viewed transnational cooperation as a means to improve its comparative position within its own national political context. The burgeoning cross-border region that these cities have helped create in the Adriatic Basin is based on the common interests of political elites more than it is on the presence of supranational cross-border projects or shared identities. Nonetheless, the cooperation that has occurred at the local level between both elected officials and NGOs has had a positive impact on both human and public security in the region as it has contributed to the reduction of social marginalization in the area. This has limited the opportunity structures for human traffickers, which has decreased levels of local clandestine migration.

This theoretical conclusion is coherent with the findings presented in Daniel Sabet's chapter on the impact of the reinforcement of the US-Mexico border on drug markets in North America. Sabet discusses the relationship between US border policies, the US-Mexico drug trade and police corruption in Northern Mexico, with particular focus on Nuevo Loredo, Tamaulipas. His chapter argues that US border security policies have created a bottleneck in transnational drug flows at the US-Mexican border. This situation has widened the opportunity structures for organized crime as the need to overcome these barriers has contributed to police corruption. Not only does such corruption facilitate the passage of illicit drugs across the border, but it also permits organized crime to act with impunity. This has been demonstrated by the increase of violence in the border region, as many activists, politicians and journalists looking to reform this broken system have been assassinated. Like Koff's chapter, Sabet argues that border security cannot be approached through the narrow policies currently in place. Instead, he argues that increased citizen participation in law enforcement oversight and expanded public discussions of security policies within civil society are the keys to successfully combating organized crime in border areas. Also similar to Koff's comparative chapter, Sabet's North American study contends that border controls are ineffective and the fight against organized crime must be linked to the improvement of representative democracy in border areas. Criminal organizations, of course, thrive in dysfunctional democracies because of existing vacuums of power and regulation which need to be filled. Thus, adequate responses must focus on the shifting nature of power within most border areas.

IV. Conclusion

The main research question to which this book responds is: "What variables most influence levels and types of integration in border regions?" Through empirical analysis of different cases in Europe and North America, the volume addresses numerous sub-themes that permeate the current literature on border studies. For example, the research presented questions why integration fails in areas where supranational initiatives are present whereas it succeeds in places where such strategies are absent (multi-level governance). Also, these studies discuss, directly or indirectly, the relationship between civil society, defined as nationalist identity, grassroots mobilization and NGO activity, and border integration (new regionalism). Finally, the book engages discussions regarding the ability and desire of leaders and citizens in border communities to distance themselves from central state authorities and create transnational territories and new forms of citizenship (the "borderless world" argument).

Most of the comparative works cited in this introduction attempt to link these sub-themes through the creation of general border theory. This approach usually examines the relationship between the three new paradigms in international relations theory. In response, the present volume adopts a very different tactic. Rather than attempting to link policy output and social outcomes in the three spheres mentioned above, the studies presented in this book open decision-making black boxes that affect each of these arenas identically. By doing-so, the volume proposes an approach that focuses on the distribution, nature and uses of power in border regions. Thus, the book connects these different spheres through the common characteristics evident in their social construction.

As stated above, the volume is based on a most different case selection. Despite their structural differences, the chapters included in this book are linked by two common traits. First, border integration is presented as a process that moves from "the bottom up" as it is created through cooperation between separate sub-national polities. In cases where local communities have no incentives for cooperation, integration is rarely furthered by supranational initiatives, even when positive historical traditions are present (see US-Mexico border).

The second similarity that permeates all of the chapters regards how politics are defined. Each of the studies adopts a systemic vision of the decision-making process which includes both elected officials and actors from civil society. These political entrepreneurs either accept or reject border integration based on short-term interests that reflect their positions in local political systems. In symmetrical border systems where actors share common goals, integration is fostered through local cooperation. When the relationship between politics on different sides of the border is asymmetrical, then cooperation is rarely viewed as advantageous.

The same can be said regarding cooperation within political communities. Decisions to support or contest border integration generally depend on the positions of political actors within local systems, whether addressing governance, civil society or economic markets. The driving force behind border integration seems to be political power, in its most basic form.

The current literature on border regions correctly argues that the nature of power has shifted in international and domestic politics. The shift from "hard" or openly coercive power toward "soft" power, based on economic influence or cultural filtration has been recognized my many authors (i.e. Nye). However, most border studies contend that these communities benefit from the blurring of the international and domestic arenas of politics which has resulted from this shift in the nature of political power.

The empirical studies presented in this volume, in fact, seem to confirm that the nature of power has indeed, changed. However, they also demonstrate that the character of political competition has not. For this reason, it cannot be considered accurate to view border politics as a simple conflict between center and periphery in which supranational initiatives improve the bargaining positions of border communities. Nor is it possible to assume that these communities automatically desire increased autonomy and that they seek more influential roles in international markets or politics.

As shown above, many observers contend that border regions are socially constructed areas. However, they do not place power games at the center of this process nor do they open this black box identified as political systems. This volume demonstrates that border politics resemble politics in all other areas. The decision-making process includes competition among political actors who generally follow their limited interests more than grandiose visions with geo-political significance. Policy output and social outcomes result from this competition. Supranational integration initiatives and the free movement of products, capital and labor (this last commodity only moves freely within the EU), have changed the opportunity structures that surround political microcosms but they have not necessarily affected power relationships within border communities. For this reason, border integration strategies enacted by the EU and the Council of Europe have had such varying effects on the ground. This approach also explains why supranational organizations have had such difficulty creating cross-border civil society, and why it has occurred in some places in North America despite the absence of structural incentives.

The nature of power seems to have changed in international politics and power in border areas reflects this shift. However, the nature of politics in these regions does not seem to have been altered despite significant changes in surrounding opportunity structures. The empirical studies included in this volume indicate that political interests, competition, and bargaining dictate modes and levels of integration in border areas. When it is not in the interests of local political elites to cooperate transnationally, in economic markets, civil society or political programs, then integration does not occur, even when the results seem counterproductive, (i.e. the fight against transnational criminal organizations). Thus, even though economic globalization and political regional integration have exposed communities to international forces, local actors still decide how much they wish to integrate into transnational and regional systems. Despite functionalist appearances, local border actors still control the permeability of national divides through both formal and informal realist decision-making. Hence, recent claims of a "borderless world" seem to be based on deceiving (dis)appearances.

References

Alper, D. 2003. "Deciphering Cascadia: Borders Unblurred", in Alper, D. and Douthit, D. eds., *Borberblur: In and Out of Place in British Columbia and the Pacific Northwest.* Bellingham: Western Washington University, p. 105-109.

Anderson, J. 2001. "Theorizing State Borders: 'Politics/Economics' and Democracy in Capitalism". *CIBR Working Papers in Border Studies* CIBR/WP01-1.

Anderson, J. and O'Dowd, L. 1999. "Borders, Border regions and Territoriality: Contradictory Meanings, Changing Significance". *Regional Studies* 33 (7), p. 593-604.

Anderson, M. 1996. *Frontiers: Territory and State Formation in the Modern World.* Cambridge: Polity Press.

Andreas, P. 2000. *Border Games: Policing the US-Mexico Divide.* Cornell: Cornell University Press.

Aykaç, A. 1994. *Transborder regionalisation: an analysis of transborder cooperation structures in Western Europe within the context of European integration and decentralisation towards regional and local governments.* Libertas.

Balme, R., ed. 1996. *Les Politiques du neo-regionalisme.* Paris, Economica.

Bauböck, R. 1997. *Citizenship and national identities in the European Union.* Cambridge: Harvard Law School.

Benington, J. and Harvey, J. 1998. "Transnational Local Authority Networking within the European Union: Passing Fashion or New Paradigm?", in Marsh, D. ed. *Comparing Policy Networks.* Buckingham: Open University Press, p. 149-167.

Blatter, J. 2001. "De-bordering the World of States: Toward a Multilevel System in Europe and a Multipolity System in North America. Insights from Border Regions". *European Journal of International Relations* 7, p. 175-209.

Bonilla, F. *et al.* eds. 1998. *Borderless Borders. US Latinos, Latin Americans, and the Paradox of Interdependence.* Philadelphia: Temple University Press.

Brooks, D. and Fox, J., eds. 2002. *Cross-Border Dialogues.* La Jolla: Center for US-Mexican Studies, UCSD.

Brunet-Jailly, E. 2005. "Theorizing Borders: An Interdisciplinary Perspective", *Geopolitics* 10, p. 633-649.

Brunet-Jailly, E. 2004. "Toward a Model of Border Studies". *Journal of Borderland Studies* 19, p. 1-18.

Caporaso, J. and Levine, D. 1992. *Theories of Political Economy.* Cambridge: Cambridge University Press.

Castells, M. 2000. *The Rise of the Network Society.* Oxford: Blackwell, 2000.

Clarkes, S. 2000. "Regional and Transnational Discourse: The Politics of Ideas and Economic Development in Cascadia". *International Journal of Economic Development* 2/3, p. 360-378.

Cornelius, W. 2004. *Controlling 'Unwanted' Immigration: Lessons from the United States, 1993-2004.* Working Paper 92, San Diego, CA: Center for Comparative Immigration Studies.

Diamanti, I. 1993. *La Lega.* Rome: Donzelli Editori.

Erie, S.P. 1999. *Toward a Trade Infrastructure Strategy for the San Diego/ Tijuana Region.* Briefing Paper, San Diego Association of Governments.

Foucher, M. 1986. *L'Invention des Frontières* Paris: La Documentation française.

Favell, A. Smith, M.P. eds. 2006. *The Human Face of Global Mobility: International Highly Skilled Migration in Europe, North America and the Asia Pacific*, New Brunswick, NJ, Transaction (Vol. 8, Comparative Urban and Community Research).

Frieden, J. 2006. *Global Capitalism: Its Rise and Fall in the Twentieth Century*. New York: W.W. Norton.

Hanson, G. 1996. *US-Mexico integration and regional economies: evidence from border-city pairs.* Working Paper 5425. Cambridge: NBER.

Herzog, L., ed. 1999. *Shared Space: Re-Thinking the US-Mexico Border Environment*. La Jolla, CA.: Center for US-Mexican Studies, UCSD.

Hix, S. 1999. *The Political System of the European Union*. New York: St. Martin's Press.

Jacobson, D. 1996. *Rights across Borders: Immigration and the Decline of Citizenship*. Baltimore: The Johns Hopkins University Press.

Joppke, C., ed. 1998. *Challenge to the Nation-State*. Oxford: Oxford University Press.

Keating, M. 2001. *Plurinational Democracy: Stateless Nations in a Post-Sovereignty Era*. Oxford: Oxford University Press.

Koff, H. 2006. "Une Comparaison des Mobilisations Politiques en France et aux États-Unis en 2005-2006". *Déviance et Société* 30 (4), p. 449-461.

Levi, M. 1988. *Of Rule and Revenue*. Berkeley: University of California Press.

Maganda, C. Dec. 2005. "Collateral Damage: How the San Diego-Imperial Valley Water Agreement Affects the Mexican Side of the Border". *The Journal of Environment & Development*, Vol. 14 (4), p. 486-506.

Marks, G. and Hooghe, L. 2001. *Multi-Level Governance and European Integration*. Boulder, Co.: Rowman & Littlefield.

Martinez, O. 1996. *US-Mexico Borderlands: Historical and Contemporary Perspectives*. Wilmington, DE: Jaguar Books.

Mattli, W. 1999. *The Logic of Regional Integration*. New York: Cambridge University Press.

Newman, D. 2006. "The lines that continue to separate us: borders in our 'borderless' world". *Progress in Human Geography* 30 (2), p. 1-19.

North, D.C. 1990. *Institutions, Institutional Change and Economic Performance*. New York: University of Cambridge Press.

Nye, J. 2004. *Soft Power*. New York: Public Affairs.

Ohmae, K. 1990. *The Borderless World*. New York: Harper Collins.

Papademetriou, D. and Meyers, D. eds. 2001. *Caught in the Middle: Border Communities in the Era of Globalization*. Washington, D.C.: Carnegie Endowment for International Peace.

Perkmann, M. 2003. "Cross-border Regions in Europe. Significance and Drivers of Regional Cross-border Co-operation". *European Urban and Regional Studies* 10(2), p. 153-171.

Perkmann, M. and Sum, N.-L. eds. 2002. *Globalization, Regionalization, and Cross-Border Regions*. Houndsmills: Palgrave.

Risse-Kappen, T. 1995. *Bringing Transnational Relations Back In: Non-State Actors, Domestic Structures, and International Relations*. Cambridge: Cambridge Studies in International relations.

Rogowski, R. 1989. *Commerce and Coalitions*. Princeton: Princeton University Press.

Sassen, S. 2000. *Cities in a World Economy*. Thousand Oaks, CA.: Pine Forge Press.

Scott, J. 1999. "Europe and North American Contexts for Cross-Border Regionalism". *Regional Studies* 33 (7), 605-617.

Williams, H. March 1999. "Mobile Capital and Transborder Labor Right Mobilization". *Politics and Society* 27, p. 139-166.

PART I

BORDERS, INSTITUTIONS AND DECISION-MAKING

CHAPTER 1

Cross-border Regionalization in an Enlarging EU

Hungarian-Austrian and German-Polish Cases

James Wesley SCOTT

I. Introduction: Border Regions as Spaces of Integration?

Border regions are spaces where nationally defined cultures, political systems, institutions and economies meet. They are also "transnational" in nature, characterised by cross-border interaction and cultural overlap, generating their own specific "Borderlands" identities. As the defensive role of state boundaries is challenged, border regions seem to be undergoing deep functional transformations. Inherent in much recent discourse on the changing significance of state boundaries is the notion that their dividing character can be overcome through the development of local transnational political communities (see Scott, 2006). Since 1989, for example, border regions have become central to European integration policies; they are understood to represent potentially flexible vehicles with which to manage conflict and facilitate collective action in the management of social, economic and environmental issues (Perkmann, 2002).

Without doubt, the European Union has played a vital role in promoting cross-border cooperation (CBC), both within the EU and at the EU's external boundaries. The principal strategy pursued by the EU in this undertaking has been to couple the development of local and regional cooperation structures with more general regional development policies. This has necessitated a process of institution-building, generally, but not exclusively, in the form of so-called Euroregions or other cross-border associations. In response to the EU's policy initiatives (and its more or less explicit institutionalization imperative) Euroregions are now present in most border regions of the EU-27, and are even developing in areas distant from the EU's borders (Zhurzhenko, 2006). However, even if the promotion of a sense of cross-border "regionness" through common institutions seems straightforward, in practice institutionalization patterns have been uneven – both in terms of governance capacities and their performance in terms of actual cooperation.

In fact, the question can be raised as to whether institutionalization as such can promote a greater sense of common region in border areas. Can the adoption of *a priori* established "models" or "best practices" promote CBC above and beyond a mere symbolic politics of "neighbourhood"? In order to address these and similar questions, I argue that "Borderlands" must be understood as products of social practices and discourses; they are complex social spaces where multiple and often asynchronous processes of regionalization unfold (see Paasi, 1999 and 2001).[1] One way to make this approach operational in a research sense is to scrutinize geopolitical rationales for CBC and their concrete reflection in regional/local cooperation agendas. In other words, the aim of formalizing CBC through rules and institutions must be set against multilevel political contexts that condition a sense of cross-border regionness. In this paper I will compare the German-Polish and Hungarian-Austrian border regions in terms of socio-political contexts for region-building and institutionalization processes that have emerged since 1990. At one level this involves geopolitical and eco-nomic rationales (e.g. traditions of interstate relations, international rapprochement and supranational integration) that frame the political symbolism of cross-border cooperation. These take expression in dis-courses and practices of CBC (including institutionalization policies promoted by the EU) that are developed within specific borderlands context. At another level, experiences of actual day-to-day cooperation

---

[1] I suggest that transboundary regionalization is perhaps best understood in terms of multilevel interrelationships between structure and agency (see Dear and Flusty, 2001). In order to comprehend the complex nature of borders and border-related identity, it is essential that these be understood as social constructs that reflect, for example, "europeanizing" and "nationalizing" influences upon cross-border inter-action as well as opportunity structures providing incentives for transboundary coo-peration. Anthony Giddens' regionalization theory has gained currency within borderlands studies, thanks largely to scholars such as Anssi Paasi (1999) and Ulf Matthiesen (2002) who have focused on the social practices and discourses involved in boundary formation. Gidden's (1984) notion of "regionalization", although not originally applied to administratively defined space as such, provides a multidimen-sional perspective for the conceptualization of region-building as a permanent pro-cess of spatial signification and "bordering". Regionalization, as understood in this abstract fashion, is a complex process of space-time zonation that is place and group-specific and that is subject to multilevel influences. Political institutions, governance principles, attitudes, local experiences, and regional identity-formation all contribute to spatial bounding and signification. Whereas internationalizing (or rather, euro-peanizing) discourses can promote an "opening" of cross-border interaction spaces, nationalizing elements can often provoke "closure" and/or ambivalence to cross-border interaction. Similarly, perceptions of interdependence and complementarity can partially suspend closure and even promote transnational behaviours.

have been influenced by local attitudes and local perceptions of borders; local perceptions of cross-border regionness have often contrasted sharply with integration rhetoric.

II. CBC and Logics of Institutionalization

The European Union has had a considerable impact on the nature of cross-border relations in Eastern and Central Europe. The EU's influence has been felt at a geopolitical level but also at a more basic societal level (Scott, 2005). On the one hand, prospective benefits of closer relations with the EU (including hopes of membership) have provided a context for rapprochement and development. On the other hand, concrete material incentives provided by the EU have been used to begin developing local and regional cooperation initiatives. In preparing Central and East European countries for membership, the EU has adopted a strategy based on institutionalized cross-border cooperation and aimed at a gradual lessening of the barrier function of national borders. These policies are also aimed at integrating previously divided border regions in order to build a more cohesive European space. Since 1990, cross-border cooperation in Central and Eastern Europe has gained momentum: contacts between local authorities have taken place regularly, and interpersonal contacts among borderland inhabitants have increased dramatically.

From an official EU standpoint, the achievement of cohesion and coherence are central goals of political integration and embodied in the 2001 White Paper on European Governance. Good internal governance and a responsive and democratic institutional architecture are, furthermore, understood to be prerequisites for promoting "change at an international level" (Commission of the European Communities, 2001, p. 26).[2] In more concrete terms this involves a process of community-building based on common rules and principles (including the so-called "acquis communautaire") as well as adherence to a comprehensive set of political and ethical values (Antonsich, 2002; Joeniemmi, 2002).[3]

---

[2] The White Paper continues along these lines with an appeal for greater geopolitical presence in order to strengthen the EU's sense of purpose: 'The objectives of peace, growth, employment and social justice pursued within the Union must also be promoted outside for them to be effectively attained at both European and global level. This responds to citizens' expectations for a powerful Union on a world stage. Successful international action reinforces European identity and the importance of shared values within the Union" (Commission of the European Communities, 2001, p. 26-27).

[3] A notable element of the Maastricht Treaty was the introduction (in Articles 8-8e) of legal and conceptual elements of formal European citizenship into an integration process hitherto characterised primarily by economic issues. Going a step further,

As a result, political exigencies of integration and enlargement as well as basic principles of EU policy, particularly structural policy, have decisively influenced the development of transboundary cooperation in Europe. Over the last decades, structures of transboundary cooperation in Europe's border regions have been built up through a combination of local initiatives and supportive measures implemented by national and European Union (EU) institutions (Kennard, 2003). This has resulted in a complex multilevel framework of formal institutions, political associations, lobbies and incentive programmes. In addition, the EU's increasing emphasis of "regionalization" and new forms of local and regional initiative, including the development of strategic alliances and "networks", are programmatic aspects of regional policy that have promoted the concept of border regions as zones of cooperation and economic "synergies" (Scott, 1999).

Cross-border cooperation in Europe is particularly in evidence in the area of regional development and spatial planning. The Council of Europe, the European Conference of Ministers Responsible for Spatial Planning (CEMAT), the Association of European Border Regions, various regional authorities and local governments in border regions as well as the European Commission itself have been deeply involved in promoting transnational cooperation in these areas. However, the most important cooperation vehicle that has emerged has been the Euroregion, a semi-formal association of local and/or regional governments, occasionally with the participation of state representatives (Scott, 2000). The rationale behind Euroregions is to establish a reliable and stable platform for cooperation in which public agencies commit themselves to developing common projects.

In its different phases of development CBC been characterised by the adaptation of existing institutional structures to new opportunities and problems set by recent geopolitical changes. Given the long track record of cross-border cooperation in Western Europe it is not surprising that cooperation stakeholders in Central and Eastern Europe have emulated many of the institutions and projects pioneered within the EU. Looking back on the history of cross-border cooperation within the EU, multilevel institutional mechanisms for transboundary cooperation in Europe appear to have contributed significantly to the development of new interregional and transnational working relationships (Perkmann, 2002). However, despite this sophisticated institutional framework, the

---

one of the implicit goals of the 1998 Treaty of Amsterdam is the promotion of a European public sphere through the establishment of common (that is unifying) constitutional principles and intergovernmental processes. These arrangements are also intended to support the definition and acceptance emergence of common values such as in the area of human rights, women's rights, democracy, etc. (Pérez Diaz, 1994).

successes of transboundary cooperation in Europe have been highly uneven.[4] In the most "successful" – that is, the most well-organized – border regions (e.g. the Dutch-German Euroregions), public-sector and NGO cooperation has been productive in many areas, especially in questions of environmental protection, local services and cultural activities. In less successful cases, cross-border projects have merely served to enhance local budgets without stimulating true cooperation. Generally speaking it has also been very difficult to stimulate private sector participation in cross-border regional development Explanations for these mixed results have been accumulated through numerous case studies, but it appears that the transcending of borders is a much more complex socio-spatial process than most empirical research has been able to capture.[5]

Given the ambiguous results of institutionalized forms of local and regional CBC within Western Europe, what can be said about the situation in the new member states (and, for that matter, at the EU's external borders)? Gabriel Popescu (2006), for example, has critically assessed EU institutionalization strategies in Central and Eastern Europe – an area of complex social, economic and political diversity. Popescu argues that Euroregions often tend to be "co-opted" by specific interests seeking either to benefit from direct EU support. As a result, Popescu states that Euroregions, especially those emerging in Central and Eastern Europe, are "top-down" creations, inhibiting processes of region-building through local initiative. If this is the general case, what then might be a more effective means of developing local and regional structures of CBC?

III. Comparing Two Regionalization Experiences

Within these contexts of border complexity the EU research project EXLINEA has analyzed how local transformation contexts interact with EU and national policies, thus conditioning cross-border regionalization projects.[6] Within the EXLINEA context, a comparison of German-Polish and Hungarian-Austrian border regions was carried in order to

---

[4]   Critical observations of the results of cross-border cooperation are provided, for example, in: European Parliament (1997), Mønnesland (1999), Notre Europe (2001) as well as in evaluations of EU structural policies such as INTERREG (http://europa.eu.int/comm/regional_policy/sources/docoffic/official/reports/p3226_ en.htm).

[5]   See, for example, Henk van Houtum's (2003) essay on "borders of comfort" and their effects on restricting cross-border economic networking.

[6]   EXLINEA ("Lines of Exclusion as Arenas of Cooperation: Reconfiguring the External Boundaries of Europe – Policies, Practices, and Perceptions") is funded through the Community Research Fifth Framework Programme of the EU, contract HPSE-CT-2002-00141. See www.exlinea.org for more information.

characterize the basic framing of CBC (and of the "common" Border-lands themselves) in terms of geopolitical rationales, cooperation dis-courses, institutionalization practices and the patterns of cooperation they have induced. As the discussion below indicates, despite the exis-tence of an overlying policy framework (i.e. the EU's CBC pro-grammes), the two regions have been characterized by rather different regionalization trajectories.

At one level, the situation facing the German-Polish and the Hungar-ian-Austrian border regions displays certain similarities. Socio-economic polarization between urban and rural areas, lack of accessibil-ity, a weak economic base in terms of innovation and high-tech sectors, large income and GDP differences and very different administrative structures are characteristic of both (Barjak and Heimpold, 2000; Bürk-ner and Matthiessen, 2002; Guz-Vetter, 2002; Jensen and Myszlivetz, 2000; Mecca Consulting, 2002). Additionally, a culture of dependence on central governments characterizes the attitudes of many local au-thorities in Poland, Hungary and other post-socialist states, thus slowing the development of local initiatives and the intensification of "horizon-tal" working relationships with other communities.

These hindrances to cooperation are compounded by large socio-economic disparities and imbalances in financial resources available to local governments in the EU and Central and Eastern European coun-tries. Conversely, looking from the EU perspective down to the local level, the policy and funding frameworks within which the two regions are developing greater internal coherence are very similar. The same rules apply and the bureaucratic procedures and complexity of manag-ing EU funds are common to both cases.

Indeed, the basic point of departure for cooperation in the German-Polish and Hungarian-Austrian border regions is challenging. And yet, both regions have developed very different institutional systems and routines for the promotion of cross-border cooperation. The reasons for this are many and defy simplistic generalization. For example, there can be no doubt that the historical and socio-cultural basis for cooperation has differed considerably in the two cases. Hungary and Austria have had an often troubled relationship but they did share an important geopolitical and economic role in Central Europe for many years as part of a dual monarchy. The relationship between Germany and Poland, by contrast, still suffers from memories of military aggression, imperialism and mutual mistrust. Furthermore, it must be added that the Hungarian-Austrian border region enjoys, in relative terms at least, more favour-able economic and demographic situation with a denser urban network – all factors that enhance cooperation.

Figure 1. The Euroregion Pro Europa Viadrina

Source: Institut für Regionalentwicklung und Strukturplanung.

What the following discussion will attempt is a comparative characterization of crossborder regionalization processes. These will be reflected, on the one hand, by specific rationales, discourses, and institutionalization processes that have shaped notion of a "cooperative" Borderlands. On the other hand, patterns of CBC that have emerged since 1990 indicate the relative success of the regionalization projects. In the interest of brevity, these case studies can only be presented in very generalized form. However, while much interesting and relevant detail cannot be touched upon, contextual reasons for diverging regionalization experience will be amply highlighted.[7]

The German-Polish Case (Region Brandenburg-Lubuskie/Euroregion Viadrina)

The region under primary consideration here is an area along the German-Polish border defined by the municipal association "Euroregion Pro Europa Viadrina (PEV)" as well as the Joint Programming Area of Brandenburg and the Voivodship Lubuskie. Administratively speaking, members of the association are counties and municipalities of the Ger-

---

[7] This case study discussion is largely based on an EXLINEA report complied by Andreas Uhrlau and James Scott (2003). This research involved extensive interviewing and document analysis during 2003.

man state of Brandenburg, and 29 municipalities of the Polish Voivod-ship Lubuskie.[8]

Cooperation Rationales

For clear historical reasons, the German-Polish relationship is a special one. Political cooperation, and most certainly CBC, has been closely intertwined with rapprochement and desire to develop a culture of mutual goodwill. As a result, much has been invested in the symbolism of binational cooperation as a response to historical traditions of conflict and prejudice. The efforts of governmental agencies of both countries, as well as the activities of many non-governmental German-Polish organizations, have provided considerable support for cross-border cooperation. At a more strategic level, the German government was also quick in acting on the political changes of 1989-1990 in Central and Eastern Europe. Sensing their future strategic economic and security importance, Germany became one of the most vocal advocates within the EU for greater regional and cross-border cooperation with these countries. Germany, among other things, took the lead in sponsoring early EU membership for the Czech Republic, Hungary and Poland.

Recognizing the need to intensify cooperation with Poland within the "pre-integration" context but also recognizing the basic structural problems of the common border region, steps were taken soon after the signing of the German-Polish Treaties to establish a variety of binational planning institutions including, at the local level, Euroregions. To this end, the EU, the Association of European Border Regions and the German Federal Ministry of Planning were instrumental in providing expertise, funds, moral support and, ultimately, in applying pressure to the Polish national government, German state governments and local communities on both sides of the border to emulate the Dutch-German model of institutionalized cooperation. As a result, a para-governmental tier of transnational cooperation has evolved along the German-Polish border (as well as along all the other borders Germany shares with neighbouring countries). While possessing no legislative and very limited executive powers above and beyond project development, these organizations of cross-border regionalism perform important coordinating governance functions. Admittedly, the "institutionalization imperative" championed by Germany fits neatly within larger European regionalization contexts which mandate the identification of regional-local gatekeepers of development resources and, at the same time, regulates the activities of subnational actors.

---

[8]　See the PEV website for information at: www.euroregion-viadrina.de or www. viadrina.org.pl.

Cooperation Discourses and Practices

As the above rationales suggest, German-Polish crossborder cooperation was framed discursively in terms of cultural understanding and
the achievement of a new quality of binational relationship. This message of atonement for past aggression informed much official cooperation at the national level, especially on the part of German state agencies.[9] At the regional level, the message of political goodwill served as
well to highlight economic development objectives. The German Land
(state) of Brandenburg was particularly active in promoting the notion
of an integrated economic space based on synergy effects (distribution
of labor and investment in industrial development). The promotion of
cross-border regional development was understood primarily in terms of
creating German-Polish production clusters as a means to stabilize
employment within a context of rapid post-socialist deindustrialization
(Gopa-BC, 1991). This was a specific preoccupation of East German
Länder which were very much concerned of being permanently rele-
gated to economic peripheries in Germany (and Europe). The Borderlands itself was conceptualised as a shared natural space in which the
Oder and Neisse Rivers, the Baltic Sea and the Ore Mountains were an
integral part. Although the notion of a common history *along* the Oder
and Neisse was evoked, there was an understandable avoidance of any
reference to pre-1945 borders in order to de-politicize the notion of a
"shared" region. Instead, the regions were conceived as spaces where
Germans and Poles might identify and pursue common interests within a
wider European context.

In terms of actual cooperation practices, the German-Polish context
has been largely influenced by public agencies and spatial planning activities. Transboundary planning cooperation was, in fact, rather productive and development concepts were drawn up at the local/regional level
during the first years of cooperation (1993-1995). These concepts embraced the ambitious objective of creating integrated economic and
ecological areas through a wide variety of measures aimed, at among
other things: combating unemployment, promoting a positive sense of
common border region identity, and fostering economic cooperation and
"good neighbourliness". In terms of spatial planning, the main actors
have been state agencies.[10] At the more local level of CBC, Euroregions

---

[9] See BMBau (1994).

[10] It is relevant to note at this point that German and Polish planning approaches were
in many ways incompatible! The traditional statist German focus on hierarchical
planning, territorial organization and central-place doctrine contrasted starkly with
the post-socialist Poland's emphasis on entrepreneurship and strategic approaches to
regional development (see Scott, 1998).

established shortly after the political changes of 1989/1990 emerged as the main vehicles for project-oriented cooperation. Through EU funding mechanisms anchored in the INTERREG and PHARE initiatives, local projects were to assume a key role in implementing ambitious schemes in these areas. However, if the experiences of the Euroregion Pro Europa Viadrina (see Figure 1) can be seen as indicative, paternalistic relations with state agencies on both sides of the border, as well as pressure from county and city governments, have basically pre-empted a clearer political role for the German-Polish Euroregions.

Regional Cooperation Patterns and Experiences

Region-building in the German-Polish case has achieved much in terms of bringing together regional stakeholders. It has also served as hands-on experience for Polish and German communities in exploiting the opportunity structures provided by the EU and helping prepare the Polish side for membership. While the direct economic benefits of cooperation have been modest, Gorzelak (2006) argues that cross-border cooperation in the German-Polish border region has had unquestionably positive results in terms of learning processes within a complex system of multilevel governance. At the least, local governments have learned to exploit and operate within EU public policy. Cross-border cooperation has empowered local governments in the German-Polish border region to act in a more forceful and self-assured manner and to grasp the potential advantages of EU integration. This has happened because they have been obliged to work with several levels of regional and national government, with different EU authorities and, ultimately, with each other. Interestingly, while Polish communities have been eligible for much less money from the EU than their German counterparts, the benefits of cooperation appear more tangible for the Polish side.

On the other hand, scholars such as Bürkner (2006) and Matthiessen (2002) have delivered less positive verdicts on German-Polish cooperation Although cross-border cooperation initiatives have flourished, spurred by the promise of EU funding for local and regional development projects, the border region remains very much divided. Apart from a few visible success stories, such as joint university facilities in Frankfurt (Oder) and Slubice and the water treatment complex in Guben/ Gubin, very little has taken place from a traditional regional development perspective. Entrepreneurial networks across the common border, for example, are weak and/or few and far between. Admittedly, the lack of progress in CBC has much to do with the development problems of the German side of the border region where high unemployment, a lack of local investment potential, depopulation and continuing marginalization continue to stymie regional development efforts (Bürkner, 2006). Indeed, specific problems of East German transformation and a greater perception of East-West contradictions have tended to limit a sense of common European purpose among the citizenry. Pessimism with regard to the EU has increased substantially on the German side, contrasting with the (more or less) enthusiastic pro-Europe stance of Polish munici-palities.

Furthermore, and at least with regard to specific planning and re-gional development priorities, political rhetoric has not translated into preferential treatment of the German-Polish border region. Incongruities between the global objectives of crossborder spatial planning, the means available for their realization and the specific priorities guiding major capital investments and regional incentives defined a difficult environ-ment for CBC in the German-Polish context. Only few areas within the common border region were viewed by the German and Polish govern-ments as being of truly strategic importance and even these received little attention.[11] The poor quality of road and rail connections between Germany and Poland, even more than 17 years since the end of state socialism, attests to a lack of binational political will to specifically promote border region development.[12]

In sum therefore, the German-Polish regional project has perhaps been a victim of its own complexity; here, institutionalization, coopera-tion principles and regional strategies came first. Effective working relationships and networks only developed later. Multilevel governance was a strategy aimed at creating avenues of communication where very few previously existed. In this respect it was a logical response to the transformations occurring in terms of both Polish-German relations and East Germany's integration into a market-oriented, federalist democ-racy. However, it at times has also given binational cooperation a somewhat ritualistic and artificial character where protocol has at times taken precedence over substantive work.[13]

---

[11] Arguably, the fatal flaw in Brandenburg's concept of cross-border economic region was an inability to mobilize the necessary financial and political resources within its own administrative machinery and thereby to send correct signals to investors and the involved public authorities. Since 1995 Brandenburg has continued to support CBC but has remained less visible on the political front.

[12] Here, attention focused on the construction and/or expansion of transport infrastruc-tures connecting Warsaw with Berlin and Paris and channelling traffic between Scandinavia and Central Europe.

[13] As confirmed in interviews with German and Polish spatial planning experts.

Figure 2. Hungarian-Austrian Border Area
(Joint Programming Area INTERREG-PHARE)

Regional structure   INTERREG IIIA - PHARE CBC: Austria - Hungary

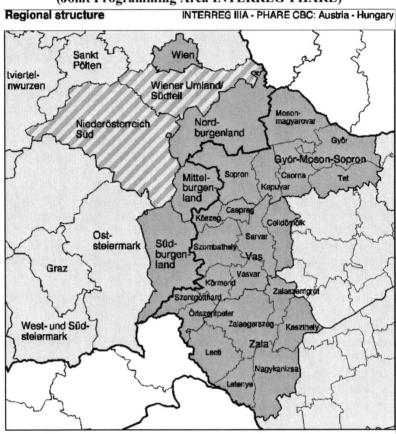

Source: Österreichisches Institut für Raumplanung.

West Pannonia: An Emerging Hungarian-Austrian Crossborder Region?

The region under scrutiny here encompasses the Austrian state (Land) of Burgenland, and the three Hungarian counties (Megye) of Győr-Moson-Sopron, Vas and Zala that make up the new Hungarian region of West Transdanubia (Nyugat Dunántúl). See Figure 2 for a more "inclusive" map that illustrates the Joint Austrian-Hungarian Programming Area for INTERREG IIIa-PHARE cbc, and thus includes the Vienna agglomeration and other areas).

Cooperation Rationales

The specific rationales behind CBC in this region can be summarized as follows: 1) the realization of regional synergy effects, 2) the strategic positioning of Hungary and Austria within the (new) European context and 3) managing social, political, environmental and economic challenges arising from EU enlargement. While, in general terms, these rationales have also characterized the German-Polish situation, the difference here is the degree of strategic engagement and political support provided by the respective national and regional levels. Planning policies aimed at developing the crossborder region cannot be seen as mere "symbolic politics" but have greatly improved – in real terms – the basis for regional synergy effects.

The border situation during the "Cold War" enjoyed strategic importance in Austrian spatial planning and regional development doctrine (see ÖROK 1975, 1978) and a great deal of conceptual work was carried out in order to formulate new development perspectives for Burgenland and other Austrian areas marginalized by the "Iron Curtain". Even before the demise of the "Iron Curtain", elements of a functional crossborder region centred around Eisenstadt, Sopron, Köszeg, communities of the Lake Neusiedler and other towns began to emerge. Crossborder commuting in the region was, for example, already a reality by the late 1980s and special border crossing passports were issued in order to expedite formalities for those affected. These patterns intensified after 1990 and were enhanced by increasing economic interaction between Vienna, Györ and Budapest. Political change provided a new basis for economic growth in Burgenland and both the national and Land governments have vigorously exploited potential for synergistic cooperation with Hungarian town and Megye to lower unemployment here. By the same token, Hungarian towns in the border region see in Burgenland (and not only the Vienna agglomeration) an important source of innovation, investment, employment and revenue.

The co-development of the Hungarian-Austrian border region has dovetailed into the overlying geopolitical and geoeconomic strategies pursued at the respective national levels. Both Austria and Hungary have aggressively pursued international economic cooperation in order to establish a commanding presence within emerging markets of Central and Eastern Europe. This has included promotion of the Vienna-Budapest-Bratislava axis as an East-West and North-South development nexus in the run-up to enlargement (see Austrian Federal Chancellery, 2000; Fath and Hunya, 1999). As a result, state agencies have successfully lobbied for improvements in the intraregional transportation system and other structural investments.

A third important rationale that has informed regional CBC between Hungary and Austria is an attempt to actively shape the political, social and regional development perspectives offered by post-socialist transformation and EU enlargement. Austria, which joined the EU in 1995 and Hungary, a new member, have a clear interest in strengthening their political voice and influence at the European level; the support of CBC between their two countries is an expression of these ambitions.

Cooperation Discourses and Practices

In terms of discursive framing, CBC in the Hungarian-Austrian case has been of a strategic and pragmatic nature and rather less idealistic. Although the pathos of 1989/1990 was not lost to the actors involved, CBC was promoted primarily as a response to new European challenges, reflecting the economic, political and environmental rationales outlines above. The principal arguments have centred on competitiveness (e.g. new opportunities for SMEs, enterprise networks, training programmes), education (regional networks of universities and polytechnics), tourism (for example, ecotourism in "binational" parklands and dialogue between communities. This discursive framing of CBC also informs the conceptualization of the Borderlands; a sense of common border region has been supported by an increasing interconnectedness of housing, jobs, and retail markets. Furthermore, advocates of CBC often refer to the border region as "West Pannonia", a name that harks back to the historical roots of the area in which the Pannonian Plain forms a general geographic frame of reference. Moreover, before the boundary changes mandated by the Paris Peace Treaties and that took place 1922, the border area was totally within Hungarian territory. This region received expression in semi-administrative terms as a Euroregion, the EuRegio West/Nyugat Pannonia, in 1999.

Opportunities for closer cross-border cooperation between Austria and Hungary began to emerge more than two decades prior to Hungary's EU accession. CBC could thus cautiously develop with the gradual improvement of bilateral relations in the 1970s and was, to an extent, supported by Hungary's reform politics during the state socialist era. Not only could the isolation of the border regions on both sides be relaxed earlier than in other central European countries, but attempts to formalize regional and local-level cooperation before 1990 also profited from the "special relationship" between the two countries. As a result, formal bilateral contacts between Austrian and Hungarian representatives began to (re)emerge in the 1980s (Hajdú, 1996).

The regional contacts that had already been established during the era of the Iron Curtain were – in what was a unique step at the time – supported by a joint Framework Programme for Cooperation towards the end of the 1980s between the Austrian state of Burgenland and the Hungarian Counties (Megye) of Győr-Moson-Sopron und Vas. This Framework Programme dealt with issues such as tourism and economic development, science, research, culture and education. At the more formal interstate level, an Hungarian-Austrian Area Development and Planning Commission (MOTTB) was created in order to facilitate intergovernmental and technical cooperation in the area of regional development.

An important milestone in the development of regional bilateral cooperation was the establishment in 1992 of a cross-border regional council that also includes the larger Hungarian cities of Győr and Sopron (these have a status similar to the Megye). This council bases its activities on the principles established by the Council of Europe's Framework Convention on Cross-Border Cooperation between Local Authorities and seeks to find more effective ways of organizing and funding projects that promote good regional relations. A further organization along these lines was established in 1998 between Burgenland and the Megyek of GMS and Vas, constituted as a "Community of Interest" (Interessensgemeinschaft) dedicated to the solution of common problems at the border. 1998 also saw the creation of a Euroregion, the EuRegio West Pannonia. Zala Megye joined this municipal organization in 1999, thus strengthening its institutional identity. The objectives of the EuRegio are to more efficiently organise cross-border cooperation through project-oriented work in the areas of spatial planning, infrastructure development, environmental protection, economic development and in social and cultural matters.

Through the INTERREG/PHARE cbc Joint Programming structures and the Small Projects fund, close cooperation between the State Government of Burgenland and the West Transdanubian Regional Development Agency has been institutionalised and regional development strategies "harmonised" for the Hungarian and Austrian subareas. Furthermore, because of stricter INTERREG requirements governing "mirrored" projects, Austrians have become more active in seeking out Hungarian partners for joint initiatives (Uhrlau and Scott, 2003).

Regional Cooperation Patterns and Experiences

In the Hungarian-Austrian case, cross-border regionalization has been privileged by context, geopolitical events and favourable economic trends. EU enlargement and the opening of borders have brought tangible gains and growth for both sides. There is also a lack of background binational conflict. While the border is unquestionably recognized as a marker of state sovereignties and diverging (now slowly converging) systemic development, research indicates that it no longer represents a hard-and-fast division in a socio-political sense and that cross-boundary relationships are part of normal everyday life. Certainly, Hungary's prospect of EU membership and the availability of EU support have helped reinforce this sentiment. However, economic complementarities and discourses of spatial-economic "common sense" are also important factors.

An important additional feature of cross-border cooperation in the Hungarian-Austrian case is the activities centred on private firms and economic actors. While the private sector is not a major player in the core "region-building" exercises promoted by INTERREG/PHARE cbc it is a fundamental contributor to regional cohesion.[14] As a result The attitude towards the region is one that is "realistic" and based on general consensus with regard to its objectives. West Pannonia is not seen as an idealistic or romantic project but rather as something pragmatic. Pragmatic definitions of region are characterised by the fact that they are more concerned with enabling individuals to act in a concerted manner in specific areas rather than predicated upon pre-defined principles. Furthermore, networking regional stakeholders and responding to local concerns is the objective, rather than the (perhaps utopian) achievement of broad socio-cultural integration. As a result, the scepticism and pessimism that pervades the German-Polish region-building project has vastly diminished here.

Despite the steady increase in cooperation intensity, the level of institutionalization and formalization of cross-border interaction remains relatively modest – particularly when compared with other border regions and the German-Polish case in particular.[15] Westpannonia is a region constituted of networks that have developed over the last two decades and that only gradually have assumed a certain institutional character or "corporate identity". Much of the cooperation that has taken place is project-oriented, involving many different actors and regional stakeholders. A Euroregion was only established in 1998, basically constituting a quasi formalization of existing working relation-

---

[14] This restriction was criticised by several interviewees.

[15] According to Jensen and Miszlivetz (2000, p. 88): "It should be mentioned that the day-to-day life shapes and forms cross-border relationships. This region is becoming more and more attractive for people from other parts of the country. Many players take the proximity of the Austrian market into consideration – the real opportunities offered by the neighbouring country. The regional economy is also important. The regional centres are becoming more diversified and are being restructured. Their contacts are becoming broader. It can be stated that Sopron, Szombathely, Kőszeg, Mosonmagyaróvár and Győr are regional centres at the border and offer more in terms of services and trade than the neighbouring Austrian towns. Therefore, contact with Hungary is important for Austrian settlements".

ships within the binational area.[16] Only in 2002 was a Transboundary Development Concept (Mecca Consulting, 2002) submitted on behalf of the EuRegion West Pannonia for a INTERREG/PHARE small projects grant. In other words, the overall development vision had time to emerge from experiences implemented over several years of bilateral project development rather than the other way around.

Presently, there are various forms of cooperation between regional institutions and authorities, especially in the field of the labor market issues, the environment, and security concerns related to the border itself across the Austro-Hungarian border. These activities are complemented by a wide variety of initiatives in social and cultural areas (e.g. such as youth exchange, local history seminars, education and vocational training) in which NGOs, non-profit organizations and institutions of higher education are the principal actors.

The region of West Pannonia has thus been developing "organically" and its future institutional shape has not been pre-determined. Furthermore, most actors involved in cooperation have indicated a preference for uncomplicated organizational structures rather than institutional complexity. This strategy has proven successful. Admittedly, however, region-building in West Pannonia has been supported by a high level of private sector cross-border activity, commuting, cross-border shopping and cultural activities.

IV. Synthesis: On the Nature of Cross-border Region-building

These case studies highlight the contextual nature of cross-border region-building. If anything has become clear in comparing the German-Polish and Hungarian-Austrian situations, it is that cross-border regionalization is inherently a process of socio-political construction and, in many, ways highly artificial. Regionalization in both cases has involved is a project of linking actor groups and institutions together with a stake in improved cooperation. By the same token, both institutional change elicited by EU enlargement and EU funding mechanisms for cross-border projects have led to a degree of "Europeanization" of the cooperation context. This is evident in the discourses, agendas and practices of cross-border actors; they very often legitimize their activities by referring to the wider political, economic and spatial contexts within which their own region must development.

---

[16]   Interestingly, the EuRegio (perhaps the most visible expression of binational "region-ness" in organizational terms) does not even have formal legal status and therefore could not directly submit project proposals to the INTERREG/PHARE cbc grants scheme.

The above discussion in no way suggests that CBC has produced rapid results in terms of economic growth and regional development. Cross-border cooperation is a process that, in these regions at the EU's former external border, can only produce long-term benefits in addressing economic and political marginality. However, the question was posed at the beginning of this essay whether it is feasible to suggest institutional models and "good" practices for cross-border region-building. Local and regional actors develop cooperation mechanisms situationally and in ways that reflect both political opportunities and social and structural constraints. Nevertheless, the results gathered within the scope of the EXLINEA research project appear to highlight the value of open-ended, project-oriented cooperation that is less rule-based.

Indeed, the story behind Westpannonia is one of pragmatic incrementalism, learning-by-doing and a gradual process of institutionalization. As working relationships have solidified, experience in joint project development has accumulated and expertise in promoting regional interests increased, as has the capacity of regional actors to take on large-scale problems and projects. In contrast, differences in East German and Polish paths of economic transformation and institutional modernization have tended to inhibit greater co-ordination of policies – even despite the "top-down" creation of several CBC institutions. Regional working relations remain weak and have received little support from the national level. And yet, despite obvious limitations, German-Polish cross-border regionalism has made considerable progress since its appearance – virtually without historical precedents – after 1989.

Finally, it remains an open question whether "Borderlands" can function as laboratories of cooperation and/or postnational political community. Cross-border regionalism does not exist for its own sake; it requires a rationale, either economic, environmental and/or political. Similarly, the removal of barriers does not guarantee a cross-border region. Only social practices and attitudes can make such a regional project reality. Liam O'Dowd (2003, p. 30) sees the European project as reconfiguring borders in terms of both "barriers" and "bridges". However, O'Dowd also acknowledges the multilevel contingency of cross-border interaction; as a result, heterogeneity is the rule and generalizations about cross-border practices are often difficult to justify. Cross-border cooperation is a very selective project of networking and "region-building". Given the simultaneity of inclusion and exclusion dynamics and discourses that characterize many borderlands contexts, the quality of cooperation will, to a great extent, depend on the role regional stakeholders and/or political elites assume in promoting a regional idea and bridging political/cultural differences (see Scott and Matzeit 2006). The quality of political messages of cross-border cooperation, however, is not only a local issue; it is subject to practices and discourses that operate at several different spatial levels and societal realms.

References

Antonsich, M. 2002. "Regionalization as a Way for Northern 'Small' Nations to be Heard in the New EU", in UPI. ed. *The New North of Europe, Policy Memos of the Finnish Institute of International Affairs (UPI)*, Helsinki.

Austrian Federal Chancellery. 2000. *Austria-Hungary INTERREG IIIa-PHARE cbc Joint Programming Document 2000-2006.* Vienna.

Baranyi, B. 2002. "Before Schengen – After Schengen. Euroregional Organisations and New Interregional Formations at the Eastern Borders of Hungary". *Discussion Papers of the Centre for Regional Studies* No. 38, Centre for Regional Studies of the Hungarian Academy of Sciences, Pécs (available from www.rkk.hu).

Barjak, F. and Heimpold, G. 2000. "Development Problems and Policies at the German Border with Poland: Regional Aspects of Trade". *European Research in Regional Science* Vol. 10.

Brown, C. and Kennard, A. 2001. "From East to West: Planning in Cross-Border and Transnational Regions". *European Spatial Research and Policy (ESRP)* Vol. 8, No. 1.

Bürkner, H.-J. und Matthiesen, U. 2002. "Sieben Thesen zu den krisenhaften Auswirkungen der Osterweiterung der Europäischen Union auf die deutsch-polnische Grenzregion". *IRS aktuell* No. 34, January.

Bürkner, Hans-Jochen and Matthiesen, U. 2002. *Grenzmillieus im potentiellen Verflechtungsraum von Polen mit Deutschland, Abschlußbericht Projekt 3.9.* Erkner-by-Berlin: Institut für Regionalentwicklung und Strukturplanung (www.irs-net.de).

Bundesministerium für Raumordnung, *Bauwesen und Städtebau (BMBau), 1994, Raumordnungskonzept für den deutsch-polnischen Grenzraum: Ziele, Leitbilder und Umsetzungsmöglichkeiten (Spatial Development Concept for the German-Polish Border Region: Goals, Visions and Possibilities for Action).* Bonn (Germany): BMBau.

Commission of the European Communities. 2001. *European Governance. A White Paper* COM (2001) 428 final. Brussels: Commission of the European Communities.

Dear, M. J. and Flusty, S. 2001. "Introduction: How to Map a Radical Break", in Dear, M. J. and Flusty, S. eds. *Spaces of Postmodernity. Readings in Human Geography.* London: Blackwell.

European Parliament. 1997. *Report on Transboundary and Inter-Regional Co-operation.* (Prepared by the Committee on Regional Policy, Rapporteur: Riita Myller), DOC_EN\RR\325\325616, Strasbourg, EU Parliament.

Fath, J. and Hunya, G. 1999. *Cross-Border Economic Co-operation on Present and future EU Borders. The Case of Austria and Hungary – A Summary of Findings.* Vienna: The Vienna Institute for International Economic Studies (WIIW).

Giddens, A. 1984. *The Constitution of Society: Outline of the Theory of Structuration.* Cambridge: Polity Press/Berkeley and Los Angeles: University of California Press.

GOPA – BC. 1991. "Förderkonzept Oder-Raum (Development Concept for the Oder Region)", Report by GOPA/Bad Homburg und BC Berlin-Consult/Berlin.

Guz-Vetter, M. 2002. *Chancen und Gefahren der EU-Osterweiterung für das deutsch-polnische Grenzgebiet.* Warsaw: Instytut Spraw Publicznych (available through www.isp.org.pl).

Hajdu, Z. 1996. "Emerging Conflict or Deepening Cooperation? The Case of the Hungarian Border Regions", in Scott, J., Sweedler, A., Ganster, P. and Eberwein, W.-D., eds. *Border Regions in Functional Transition. European and North American Perspectives.* Regio Series of the IRS, Vol. 9: Erkner by Berlin, IRS.

Jensen, J. and Miszlivetz, F. 2000. *"Preparity" – Structural Policy and Regional Planning Along the External EU Frontier to Central Europe: Preparity for Eastern Enlargement* (see www.euregio.hu/ujterv).

Joenniemi, P. 2002. *Can Europe be Told from the North? Tapping into the EU's Northern Dimension.* Working Papers of the Copenhagen Peace Research Institute, No. 12/2002 (Contact through www.copri.dk).

Kennard, A. 2003. "The Institutionalization of Borders in Central and Eastern Europe: A Means to What End?", in Berg, E. and van Houtim, H. eds. *Routing Borders Between Territories, Discourses and Practices.* Aldershot: Ashgate.

Matthiesen, U. 2002. "Transformational Pathways and Institutional Capacity Building: The Case of the German-Polish Twin City Guben/Gubin", in Cars, G., Healey, P., Madanipour, A. and de Magalhaes, C., eds. *Urban Governance, Institutional Capacity and Social Milieux.* Aldershot: Ashgate.

Martinez, O. 1994. "The Dynamics of Border Interaction: new Approaches to Border Analysis", in Schofield, C., ed. *Global Boundaries. World Boundaries* Vol. 1: London and New York: Routledge.

Mecca Environmental Consulting. 2002. *Entwicklungsleitbildes „Burgenland – Westungarn" Ein Programm zur Entwicklung der EuRegio West/Nyugat Pannonia*: Eisenstadt/Sopron: Mecca Consulating TH Eisenstadt.

Mønnesland, J. 1999. "Cross Border Programmes versus Regional Development Programmes. The population Density Influence": *Paper presented at the Regional Studies Association Conference*, 21-24 September, Bilbao, Spain.

Notre Europe – Groupement d'Études et de Recherches. 2001. *Is the New Europe Inventing Itself in Its Margins? Cross-Border and Transnational Cooperation* (available through www. notre-europe.assoc.fr).

O'Dowd, L. 2003. "The Changing Significance of European Borders". *Regional and Federal Studies* 12 (4), p. 13-36.

Österreichische Raumordnungskonferenz-Örok. 1975. *Die Grenzgebiete Österreichs.* ÖROK-Schriftenreihe, No. 7. Vienna: ÖROK.

Österreichische Raumordnungskonferenz-Örok. 1978. *ÖROK-Regionalpolitik in Ost-Grenzgebieten.* Vienna: ÖROK.

Paasi, A. 1999. "Borders as Social Practice and Discourse: The Finnish-Russian Border". *Regional Studies* Vol. 33, No. 7.

Paasi, A. 2001. "A Borderless World" Is it Only Rhetoric or will Boundaries Disappear in the Globalizing World?", in Reuber, P. and Wolkersdorfer, G. eds., *Politische Geographie. Handlungsorientierte Ansätze und Critical Geopolitics.* Heidelberg: University of Heidelberg (Heidelberger Geographische Arbeiten).

Perkmann, M. 2002. "The Rise of the Euroregion. A Bird's Eye Perspective on European Cross-Border Co-operation", published by the Department of Sociology, Lancaster University (www.comp.lancs.ac.uk/sociology).

Popescu, G. 2006. "Geopolitics of Scale and Cross-Border Cooperation in Eastern Europe: The Case of the Romanian-Ukrainian-Moldovan Borderlands", in Scott, J., ed. *EU Enlargement, Region Building and Sifting Borders of Inclusion and Exclusion.* Aldershot: Ashgate.

Scott, J. 1998. "Planning Co-operation and Transboundary regionalism – Implementing European Border Region Policies in the German-Polish Context". *Environment and Planning C: Government and Policy* 16 (5).

Scott, J. 1999. "European and North American Contexts for Cross-Border Regionalism". *Regional Studies* Vol. 33, No. 7.

Scott, J. 2000. "Euroregions, Governance and Transborder Co-operation within the EU". *European Research in Regional Science* 10.

Scott, J. 2005. "The EU and 'Wider Europe': Toward an Alternative Geopolitics of Regional Cooperation?". *Geopolitics* 10 (3).

Scott, J., ed. 2006. *EU Enlargement, Region Building and Sifting Borders of Inclusion and Exclusion.* Aldershot: Ashgate,

Scott, J. and Matzeit, S. 2006. *EXLINEA Final Project Report*, available at: www.exlinea.org.

Uhrlau, A. and Scott, J. 2003. *Background Case Studies of the Hungarian-Austrian and German-Polish Border Region, Workpackage 3 of the EXLINEA project*, October 2003 (www.exlinea.org).

van Houtum, H. and Van der Velde, M. 2003. "The Power of Cross-border Labour Market Immobility". *Regional and Federal Studies* Vol. 12.

Zhurzhenko, T. 2006. "Regional Cooperation in the Ukrainian-Russian Borderlands: Wider Europe or Post-Soviet Integration?", in Scott, J., ed. *EU Enlargement, Region Building and Sifting Borders of Inclusion and Exclusion.* Aldershot: Ashgate.

CHAPTER 2

The Sub-national Aspects
of the EU-RF Relationship

Cross-border Cooperation in the Context
of Europeanization and Democratization

Anastassia OBYDENKOVA

I. Introduction

Strengthening democracy in Europe is a central objective of the European Union. Similarly, Russia has been attempting to improve democratic practice domestically. The process of democratization includes both national and supranational influences. It can also be analyzed on both the national and sub-national (regional) levels. This chapter aims to identify the factors that determine the development of cross-border cooperation (CBC) between the regions of Russia and the EU. By doing so, it also responds to the following research question: How has the CBC influenced democratization on the sub-national level, in the regions of a EU-neighboring country, such as Russia?

Within the current literature on democratization in Russia, little clarity has been achieved regarding the impact of the EU on this process. This chapter aims to show that the analysis of CBC can help to understand the importance of the EU in democratization processes. Through the study of regional development in Saint-Petersburg and Leningrad oblast, I argue that levels of regional involvement in cross-border cooperation with Europe and democratic performance vary sub-nationally. To sum up, this chapter analyzes CBC in the context of Europeanization and Democratization. It aims to provide some evidence regarding CBC and, thus, to contribute in responding the main question of this book: "What variables most influence levels and types of integration in border regions?" (Koff, 2007, p. 17).

During the period of regime transition in the 1990s, the regions, or constituent units (CUs), of Russia were given extended autonomy by the central government. However, only some of the CUs used this autonomy to establish their own foreign policy with European actors (regions and organizations of Europe). This presented a puzzle: why did only some of the regions profit from increased autonomy that was granted to most of the CUs. Not all of the regions were active in establishing their own foreign relations. Why did some of the regions opt to act on the international level while the others were reluctant to undertake such initiatives? What factors, apart from geopolitical ones, have encouraged cross-border cooperation and regional participation in international and specifically, European, affairs?

From another perspective, this analytical puzzle opens a different but equally important question: Has there ever been any significant impact of cross-border cooperation on democratization at the sub-national level? Russia's regions, in fact, provide an excellent opportunity for a comparative study of policy impact since many key non-policy variables remain constant (history, post-Soviet political culture, institutional legacies and international environment).

In order to address both of these challenging questions, this research analyzes CBC as both a dependent and an independent variable. Above all, it analyzes CBC in the context of "Europeanization" and "Democratization".[1] On the one hand, it applies this concept *beyond* the EU, investigating possible impacts of the EU on non-members and non-candidates. This study geographically extends the notion of Europeanization beyond the EU's formal borders and it limits its functional interpretation. In other words, Europeanization describes the democratic impact of the EU through cooperation and value expansion on "smaller" partners, such as the regions of the RF. The research narrows the interpretation of Europeanization to:

(1) the processes and mechanisms by which European institution-building may cause change at the domestic level;

(2) the development of networks of interactions among domestic and supranational actors (in this study, between the regions of a non-EU state and the EU and the regions of the EU-states); and

(3) the gradual and differentiated diffusion-penetration of democratic values, general democratic norms from those European institutions into the domestic politics of the regions of Russian Federation.

An important point concerning operationalization of the concept of Europeanization is the absence of "pressures for adaptation" in all of the

---

[1] Theoretical analysis of Europeanization is given in detail in Morlino, Leonardo. 2002. "The Europeanisation of Southern Europe", in Costa Pinto, A. and Teixera, N.S. eds. *Southern Europe and the Making of the European Union 1945-1980*. New York: Columbia University Press, p. 237-260; and Obydenkova, A. 2006. "Democratization, Europeanization and Regionalization beyond the European Union: Search for Empirical Evidence". *European Integration online Papers* Vol. 10, No. 1, 2006.

case-studies. The regions of the RF are not "forced" by the central government to develop cooperation links with European neighbors but, on the contrary, they are sometimes restricted in developing such links. Thus, this major difference between regions of the EU-members and candidates allows us to investigate Europeanization as a dependent variable as well.[2] It allows us to ask the question: What factors "encourage" certain regions to develop cooperation with actors already-integrated in the Europe Union?

Almost every constitution and charter adopted by the regional governments in the 1990s proclaimed the "sovereign, democratic, rule-of-law state" and gave a list of civic, economic, and political freedoms: the division of powers; rule of law; the invalidity of secret laws; open, free, and fair multi-party elections; basic freedoms of speech, assembly and conscience, the defence of all forms of property rights, etc. However, in some of the regions, such as Kalmikia, Primorskii krai, Tatarstan, etc., behind the formally established democratic "rules of the game", one would fine almost no division of powers, manipulation of elections, suppressed freedoms of speech, etc.

Responses to the aforementioned questions vary significantly. The two predominant approaches utilized address

1) "contextual" factors (the role of geopolitical location and the length of the border) (Hypothesis 1) and

2) the impact of domestic policy factors within Russia (such as the process of regionalization and federalization in the 1990s) (Hypothesis 2).

According to the former, border location is a significant variable. For example, we can hypothesize that regions located on the Northwest border of Russia are more active in the establishment of cross-border regional cooperation with European neighbors. Consistent with Karl Deutsch's theory of regional integration, we may suggest that the longer the border with the EU, the more a region would be open to cooperation.[3] Thus, the borders are viewed not as "dividing lines" but as "point of cooperation".

---

[2]   The theoretical analysis of Europeanization as an independent variables is presented in: Obydenkova, A. 2005. "Puzzles of the European Regional Integration and Cooperation: Interplay of 'Internal' and 'External' Factors", in Di Quirico, Roberto ed. 2005. *Europeanisation and Democratisation. Institutional adaptation, Conditionality and Democratisation in European Union's Neighbor Countries*. European Press Academic Publishing, p. 199-222.

[3]   Deutsch, Karl W. 1953. *Nationalism and Social Communication. An Inquiry into the Foundations of Nationality*. The M.I.T. Press; and Deutsch, Karl W. 1966. *National-*

The second hypothesis suggests that domestic policy variables, such as asymmetrical federalism (subdivided into constitutional and contractual asymmetry) might have influenced the initiatives of the regions towards Europe. During the 1990s, some of the regions of the RF received greater autonomy due to stipulations in the Federal Treaty (1992) and the RF Constitution (1993). This phenomenon was described as constitutional federal asymmetry. About 50% of the regions have signed power-sharing contracts with the central government outlining additional autonomy. Thus, we hypothesize that both groups of regions had more "institutional space" for maneuvers on the international arena.

In addition to the aforementioned approaches, an alternative hypothesis contends that regions of a transitional state, interacting with regions of established democratic states through participation in different EU programmes aimed at nurturing democratic values, would be more democratic than the others (Hypothesis 3). This alternative strategy examines the impact of external, supra-national factors on democratization processes.

On the one hand, we have a transnational actor, the EU, which is composed of well-entrenched democracies, and on the other hand, its biggest neighbor, the RF, is a "transitional" state. Political regimes that existed in Russia from the mid-1950s until the late 1980s – both at national and regional levels – were commonly regarded as authoritarian. There were some differences in the relative economic development and ethnic composition of Russia's administrative units but the regional regimes were still similarly configured along the lines of a set of actors and institutions. However, in the late 1990s, the varieties of political regimes in Russia demonstrated large-scale diversity in their regional politics. Regimes with some features of democracy blossomed in St. Petersburg, while authoritarianism thrived in Kalmikiya, as did "warlordism" in the Primorskii *krai.* Some hybrid regimes were found in other regions.

Thus, this chapter examines whether the regions of Russia involved in cooperation with European regions and organizations would be more pro-democratically developed than the others. This analysis would allow for the assessment of the impact of Europeanization on democratization through the comparison of two regions with similar characteristics, St. Petersburg and Leningrad oblast. The findings have implications for the literature on new regionalism, international relations and globalization studies: the emergence of "new" regions through transna-

ism and Social Communication. Cambridge, Massachusetts, and London, England: The M.I.T. Press, Massachusetts Institute of Technology.

tional cooperation analyzed as both a dependent and independent variable.[4]

This chapter will proceed with a brief overview of the EU policies towards democratic development in Russia, including a description of their mechanisms and implementing institutions (section II). Part three presents a brief description of "contextual" and institutional factors as factors influencing the development of CBC and it focuses on the comparative analysis of St. Petersburg and Leningrad oblast across these variables. Part four then examines the impact (if any) of cross-border cooperation on the process of sub-national democratization within the selected regions. Following the general logic of this chapter, the conclusion, section five, has two main objectives. It aims to clarify both the role of contextual and institutional factors on the development of CBC with Europe as well as the impact of this cooperation on democratization in the regions. Through the arguments presented below, it addresses the three competing hypotheses previously described.

II. Cross-border Cooperation in the Context of Europeanization:

The EU's Policy as an External Lever of Democratization

The common border between Russia and the EU (Finland, Estonia, Poland, Latvia, and Lithuania) is often viewed as a point of further integration (cultural and economic) rather than as a "new dividing line". Apparently, the EU is interested in strengthening its northern border and the north-western regions of Russia constitute a significant part of this area. Therefore, the EU policy in this region is in fact an "external" factor that influences the process of regionalization within Russia. However, the phenomenon has a rather positive connotation. It is viewed as a lever of regional democratization, which has been implemented through numerous EU programmes launched in this geographical area. Thus, it seems that the EU does have some impact on regime transition in a number of regions, which permits us to utilize the phrase: "Europeanization of the regions" (given that "Europeanization" is defined as democratic impact through value expansion and networking). Apparently, the Europeanization of the regions of Russia is not as strong as that which occurs in EU member and candidate-states. However, it is a significant factor within the framework of cross-regional analysis in

---

[4]   On "new regionalism", see Obydenkova, A. 2006. "New Regionalism and Regional Integration: The Role of National Institutions". *Cambridge Review of International Affairs* Vol. 19, No. 4 (December 2006).

the RF. This section looks at the instruments and institutions employed by the EU in its approach to the regions of Russia.

By 1999, in terms of treaties, the European Community (EC) had only the now outdated Trade and Cooperation Agreement (TCA) with Russia as it was signed with the USSR in December 21, 1989. The only real instrument of politics towards Russia was the TACIS programme initiated in 1991. The first attempt at giving strategic direction to the evolving EC-Russia relationship was the decision to negotiate a Partnership and Cooperation Agreement (PCA) in March 1992. The PCA still remains the legal basis for the relations between the EU and Russia. Respect for democracy and human rights has been elevated as the leading principle of the future development of this partnership. "Cooperation", "partnership", "involvement" are the key-words defining the mode of "integration" of Russia and its regions into Europe and they are used interchangeably.

The preamble to the agreement stresses the importance of the "common values" shared by the EU and Russia – respect for human rights, the rule of law, and the market economy. The fulfillment of the "obligations under the Agreement", as conditions for cooperation are interpreted as a direct reference to the primacy of the observance of democracy and human rights in Russia.[5] Therefore, from the very outset of the EU-Russia relationship, cooperation was based on democratic values considered to be "common values" for both sides.

The first more concrete strategy of the EU towards the RF which was based on the European Commission's communication,[6] was presented at the meeting of the EU foreign ministers in Carcassonne in March 1995. The main aim of the strategy was to develop a mutually beneficial partnership with Russia, based on mutual responsiveness and respect for human rights. The communication declared the following aims: strengthening political, societal and economic stability in Russia, and sustainable development, which will improve the living standards of the Russian population, and increase cooperation in resolving the most important regional and international questions.

The preparation for the drafting of the first Common Strategy on Russia was started in the aftermath of the financial and political crisis of 1998. The draft was presented at the Cologne European Council. The heads of state and government then formally adopted the first Common

---

[5] Cf. Timmermann. 1996. "Relations between the EU and Russia: The Agreement on Partnership and Co-operation". *Journal of Communist Studies and Transition Politics* 12, p. 196-223, p. 224.

[6] Commission of the European Communities. "The European Union and Russia: the future relationship". COM (95) 223 final, 31 May 1995

Strategy on Russia (CSR). The CSR declared two strategic goals concerning Russia: "a stable, open and pluralistic democracy, governed by the rule of law and underpinning a prosperous market economy" and "maintaining European stability". The four principal objectives of the CSR are: (1) consolidation of democracy, the rule of law and public institutions; (2) *integration* of Russia into a common European economic and social space; (3) cooperation to strengthen stability and security in Europe and beyond, and (4) common challenges on the European continent.[7]

The discourse had changed over time, as these documents show. It starts with such notions as "cooperation", "partnership", "involvement", and gradually, in CSR, came to use the notion of "integration". Based on the general guidance of "integration", the document spells out the numerous "areas of action" where further measures are required, such as developing training programmes for Russian civil servants, promoting cultural and educational exchanges, conducting high-level dialogue on economic issues, working with Russia to develop a joint foreign policy initiative, and overall environmental protection. Once these general principles were in place, the member-states could then add their own proposals. Thus, Finland entered a reference to the Northern Dimension; Sweden added the importance of free media; Germany insisted on the importance of economic reforms in Russia and Great Britain was interested in nuclear safety. The CRS as a part of a learning process as a way of approaching "common values and fundamental interests".[8]

The "Common Strategy" also focuses on "cooperation to strengthen stability and security in Europe and beyond". It addresses "common challenges on the European continent" – social and environmental dangers, organized crime (money laundering, trafficking in drugs and people); illegal immigration; and terrorism. There is enormous scope for cooperation in meeting these challenges. There is already joint action to fight organized crime, involving cooperation between EU and Russian law enforcement agencies, and several important environmental projects, especially in northwestern Russia under the "Northern Dimension" initiative.

---

[7]   Haukkala, H. 2000. "The Making of the European Union's Common Strategy on Russia", in Haukkala, H. and Medvedev, S. eds. *The European Common Strategy on Russia. Learning the Grammar of the CFSP.* Helsinki: The Finnish Institute of International Affairs.

[8]   TEU Article 11.1. Cited in Haukkala, H. 2001. "The Making of the European Union's Common Strategy on Russia" in Haukkala, H. and Medvedev, S. eds. *The European Common Strategy on Russia. Learning the Grammar of the CFSP.* Helsinki: The Finnish Institute of International Affairs, p. 66.

TACIS is the largest technical assistance programme in Russia. This programme is intended to facilitate the transfer of western "know-how" and expertise to assist in the development of the institutions, legal and administrative systems, management skills essential for a stable democracy and a properly functioning market economy. An "indicative programme", covering four years at a time, provides a policy framework for the operation of TACIS in Russia, and identifies three crucial areas: support for institutional, legal, and administrative reform; support to the private sector and assistance for economic development; and support in addressing the social consequences of transition. Most of the training projects have been targeted at civil servants and local government officials, judicial and law-enforcement personnel, and discharged military officers in some of the regions. Twinning projects facilitated the exchange of experience and the encouragement of networking is increasingly seen as a vital part of many TACIS initiatives. The TACIS Tempus programme has encouraged universities in EU member states to form partnerships with their counterparts in Russia, in order to stimulate reform in higher education, and to facilitate the mobility of staff and students. There has also been a distinct TACIS Democracy Programme to promote democratic values and practices throughout Russian regions.

The Northern Dimension (ND)

Although the "ND" is not exclusively directed at Russia, it provides opportunities for constructive engagement and integration of separate regions of the RF into European political and cultural life. It is the result of an initiative in 1997, sponsored by Finland, to encourage closer cooperation among all states and regions in northern Europe, irrespective of whether they are EU members or not. The ND was approved at the European Council in Vienna in December 1998 and formally launched the following year at the Council in Helsinki. In the context of European integration, the overriding objective is to encourage people and institutions in northwestern regions of Russia to feel that their homeland forms an integral part of the region, rather than being isolated and potentially, therefore, alienated.

The ND does not involve either new institutions or financial instruments. One of the most important principles is "positive interdependency" between the EU, the Baltic Sea region and Russia, and the objective is to ensure "win-win" outcomes from concrete projects that bring clear benefits both to Russia and to its regional neighbors. An "Action Plan" identifies a large number of areas in which crossborder cooperation on concrete projects would be beneficial. These include transport, energy, nuclear safety, the environment, public health, trade, international crime, etc. All specific actions, especially those that involve finance, have to be undertaken through existing legal and financial instruments (PHARE, TACIS, and Interreg) or with the support of other international organizations, such as the European Bank for Reconstruction and Development of the Nordic Investment Bank.

The initiative focuses on relations between Finland and Northwestern Russia. It started with the restoration of cooperation, especially in economy and trade, but gradually this idea has grown into a proposal for large-scale cooperation, including not only the EU and Russia, but also the Baltic states.

The Involvement of Russian Regions in Baltic Sea Cooperation

The Council of Baltic Sea States (CBSS) and the Union of the Baltic Cities (UBC) could become bodies which will be helpful for cooperation and negotiations. All the countries of the Baltic Sea region are members of the Council. The Union of Baltic Cities includes almost 100 cities of the Baltic Sea region. The organization plays a positive role in developing ties on a sub-regional level. Although UBC is not an organization of high political significance, it could help solve practical problems and could increase of cooperation.[9]

III. Contextual and Institutional Factors of Cross-border Cooperation

St. Petersburg and Leningrad Oblast in the 1990s

Contextual Factors

Regionalization in Europe has been accompanied by regionalization within Russia itself, which is based on a number of so-called "contextual" and "domestic-institutional" factors. The country is geographically divided into "European" and "Asian" parts, and regions also differ in a geopolitical sense, in ethnic composition, and in levels of economic development. What factors might help to initiate cross-border communication and cooperation between the regions of Russia and the EU? These factors can be conditionally divided into "contextual" (geopolitical, ethnic, and economic) and institutional (constitutional and contrac-

---

[9] See for example, Kivikari, Urpo. 1998. "The Application of growth triangle as a Means of Development for the Kaliningrad region", in Kivikari, U., Lindstrom, M., and Liuhto, L. eds. *The External Economic relations of the Kaliningrad region.* Turku: School of Economics and Business Administration, Institute for East-West Trade C2: 1-13 and Khudoley, K. "Russian-Baltic Relations – a View from Saint Petersburg", in Hubel, H. ed. *EU Enlargement and Beyond: The Baltic States and Russia.* Berlin: Verlag Arno Spitz.

tual federal asymmetries) variables.[10] Among the geopolitical conditions are the existence (or absence) of common border with the EU, the length of the border as an interaction point, the location in either the "Asian" or "European" part of the country, or location in the Northwest part of Russia.

The position of the titular ethnic group in many CUs is quite weak, compared with the other national groups in these areas. The ethnic groups are highly dispersed across the territory of the RF because of the immigration policies of the tsarist period (especially under the rule of Catherine II) and the Soviet era (most notably during Stalin's rule). According to the 1989 census,[11] the titular nation made up less than half of the population in fourteen of the administrative units that are RF republics today. In Kabaradino-Balkaria and Dagestan, a majority exists only if two or more titular groups are added together. It leaves only five republics in which a singular titular nation forms the majority of the population – Chuvashia, Tyva, North Ossetia and Chechnya, and Ingushetia.

In autonomous oblasts and autonomous okrugs (which have a lesser degree of autonomy than the republics) the presence of members of the titular nation is even less. Thus, for example, in the autonomous okrug of Khanti-Mansi, the two titular groups together account for no more than 1.4% of the total population of this CU. In general, the proportion of the titular nations in these units is quite low.[12]

This factor can be viewed in terms of economic dependence, rather than interdependence. Many of the ethnically defined units had developed a dependence on the center during the Soviet period. The local economies functioned as integrated parts of the Soviet economy. Planning and investment were always carried out within the confines of a region, for a particular industry; without developing a balanced, self-sufficient economy within the republic or okrug.

---

[10] For more detailed analysis of geopolitics, ethnicity, and economic development of the regions of Russia see Obydenkova, A. 2004. "The Role of Asymmetrical Federalism in Ethnic-Territorial Conflicts". *Working Paper of the European University Institute*. Florence, Italy: European University Institute, SPS 2004/16.

[11] Census 1989 of the RF. Moscow. My purpose was to find indicators for the very outset of the transitional period, i.e. the 1990s. This was the initial reality that politicians at the national and regional level were faced with when the USSR collapsed. Thus, the census of 1989 was ideal for the purpose of this research. The later census of 2002 could not be used for this particular study, as it reflects the changes which took place after 12 years of transition and, thus, presents data inapplicable for testing the hypotheses.

[12] The Komi-Permiak autonomous okrug and the two Buryat-inhabited okrugs where the share of the titular nation did not surpass 17% might be considered exceptions.

The areas where there is the greatest potential, for the development of a more or less independent, economy are the Volga-Ural area and northern Siberia – with their rich deposits of oil, gas and other natural resources. The republics of the northern Caucasus are among the poorest and the least developed CUs. The republics of southern Siberia are also highly dependent on transfers of federal funds.[13] Most of the republics can be defined as "mono-economies", in the sense that they rely on imports from other parts of the Federation. For example, 80% of these goods sold in the republics were imported from former union republics.

To sum up, given great disparities in the level of economic development, ethnic composition, and geopolitical location, federalism seems to be the only feasible option to accommodate such differing regions. Above all, it is the asymmetrical federal arrangement which offers the "individual" approach to managing the center-periphery relationship in a multi-ethnic state. Thus, given numerous ethnic groups, geopolitical and economic disparities, it seems that the establishment of certain institutional mechanisms in the form of federal asymmetry is almost unavoidable.

Institutional Factors: Asymmetrical Federalism

Soviet Russia was a highly centralized state. However, the beginning of the 1990s signaled the start of critical changes not only on the national level (in the framework of regime transition) but also on the regional level (through decentralization reforms which took the form of asymmetrical federalism). In 1991, the RSFSR's administrative-territorial structure was modified and this change was later codified in the Federation Treaty of March 1992 and the Constitution of 1993.[14] The 16 autonomous republics, and 4 of the 5 autonomous oblasts, were given the status of "republics". The other 68 CUs (including 49 oblasts, 7 krais, 2 federal cities, 1 autonomous oblast, and 10 autonomous okrugs) became known as "regions" of the RF.

In addition to the Federal Treaty, President Yeltsin signed three other treaties in March of 1992: one with the autonomous republics and the autonomous oblasts that elevated them to the status of a republic (these are Adygeia, Gorno-Altai, Karachay-Cherkessia, and Khakassia); one treaty with autonomous okrugs; and another treaty with non-ethnic oblasts, krais, and the two cities of Moscow and St. Petersburg (which

---

[13] The best example of it is the fact that 90% (!) of expenditures in the Tyvanian budget have been covered by federal subsidies.

[14] Obydenkova, A. 2005. "Institutional tools of conflict management: Asymmetrical Federalism in Ethnic-Territorial Conflicts: Quantitative Analysis of Russian Regions". *Peace, Conflict and Development: An Interdisciplinary Journal* Issue 7, July 2005.

received the status of federal cities which made them equal to an oblast). The Federation Treaty described "republics" as "sovereign states" implying extended rights for this group of CUs in the areas of natural resources, external trade, and internal budgets. Tatarstan and Chechnya refused to sign the Federal Treaty, seeking the more clearly defined status of independent states. All other CUs, apart from the republics, secured enhanced rights.[15]

The Federal Treaty completed the construction of "constitutional asymmetry" (as it became a part of the new RF Constitution). It described republics as "sovereign", which suggested that the republics not only had a right to refuse to join the federation, but also could secede of their own initiative.

The 1993 Constitution took precedence over the Federal Treaty. In drafting Russia's constitution, Yeltsin insisted on three principles: human rights were to be guaranteed throughout Russia (including the republics); the unity of the RF must be maintained; the constitutions of the republics should not contradict the Russian constitution. The definition of the republics as "sovereign states" was dropped, while the federation structure still included different approaches to CUs. The Constitution declared (Art. 5) the equality of all subjects of the Federation, when in reality they were entitled not only to a different status but also to different rights. One of the most striking differences was that the republics were granted all the attributes of a sovereign state (constitutions, presidents, legislature, etc.) while all other CUs were granted the right to have charters, governors, and more stringent tax payments.[16] Additionally, all CUs were divided into "ethnic regions" (republics, autonomous oblast, autonomous krais) and "territorial regions" (oblasts and krais).There are 32 CUs defined as "ethnic regions". This group includes 21 republics, 10 autonomous okrugs and 1 autonomous oblast.

The "constitutional asymmetry" was followed by "contractual asymmetry". In February 1994, President Yeltsin signed the bilateral treaty with Tatarstan. In the signing of this treaty, Yeltsin encouraged other CUs to follow suit. By 1996, similar treaties were signed with Kabardino-Balkaria, Bashkortostan, North-Ossetia, Sakha, Buryatiya, Udmurtia. In 1996, similar treatment was accorded to Sverdlovsk, Orenburg, Kaliningrad, Khabarovsk, and Komi. These bilateral treaties (also called "power-sharing agreements") helped to resolve some of the tensions between the federal centre and the regions. In addition, they

---

[15] On asymmetric federalism and de-centralization in the 1990s, see for example Kahn, J. 2000. *Federalism, Democratization, and the Rule of Law*; Ross. 2000. *Federalism and Democratization in Russia*.

[16] *Ibid.*

gave sufficient autonomy to the administrations of the regions to rule their domestic policy and often some gave certain concessions for the conduct their own foreign policy. By the end of the 1990s, about 50% of all regions had signed power-sharing agreements with the central government in Moscow. These treaties (or contract) normally outlined the "extra-autonomy" the regions have received in domestic and foreign policy areas. On the other hand, it has created an extremely asymmetrical federal arrangement by "privileging" some regions over the others through the signing of bilateral power-sharing agreements. This phenomenon is conditionally labelled as "contractual asymmetry".

Many regions of the RF have asked for extended autonomy and were granted either constitutional and/or contractual autonomy. Thus, it would be natural to suggest that the regions which had received higher autonomy through bilateral power-sharing agreements with the central government had more opportunity to develop cross-border cooperation with European partners.

However, apart from institutional factors, geopolitics may also provide potential explanation. The regions located in the north-west of Russia, and especially those regions which share borders with EU-countries, have more chances to engage in different programmes launched by the EU (under the auspices of Northern Dimension). They present a geopolitical interest for the EU states and regions. Location on periphery of the RF makes them closer to the EU than to Moscow. The development of cross-border regional communication and cooperation are often viewed as initial steps towards broadly defined integration on the regional level which already has taken the form of Euroregions.

Given these two conditions, geopolitics and institutions, I present two regions as case-studies to control for both geopolitical location and for degree of autonomy. We have chosen two regions with a similar degree of autonomy and a similar location, to see whether these two factors – geopolitics and autonomy – are sufficient for the development of cross-border cooperation and communication with Europe.

In this particular chapter, "cooperation" is defined as involvement in different political projects (as, for example, environmental projects) and regular trade (which was developed on a regional level during the 1990s). "Communication" is a broader notion which in this research includes those exchange of opinions which takes place during the process of cooperation, as it is described above, *and* "communication" which takes the form of regular inter-regional academic exchanges, expert exchanges, learning/teaching programmes, and regular inter-regional conferences.

First, it is apparent that geopolitical location is one of the crucial factors in initiating the para-diplomatic activities of the regions towards the EU and its regions. Second, it is also apparent that the CUs with enhanced autonomy have more institutional "space" to develop cross-border cooperation with Europe. The question is what other factors, apart from these "apparent" explanations, may hinder the ability and motivations of the regions to develop such cooperation. By examining two regions with equally favorable geopolitical location and equal autonomy from the central government, this comparison allows for the identification of other potential explanatory variables.

The third question is what is the impact of cross-border cooperation with Europe on the regime transition in the regions involved in regular cooperation and communication? What is the *regional* impact of Europeanization? To answer these questions, the paper focuses on the regions with the same geopolitical location and, then, it analyzes the level of democratization within these regions as one of the trends of Europeanization. We focus on two regions located on the Northwest border of the RF.

Given that both regions share the border with the EU, we control for the geopolitical variable, which facilitates the identification of other hidden factors that might have influenced the activities of the regions toward Europe.

Have they used their preferable geopolitical position to develop cross-border cooperation with EU countries and regions? Are they equally involved in the numerous programmes launched by the EU under the auspices of Northern Dimension? These are the questions to be answered.

St. Petersburg and Leningrad Oblast

Geopolitics

Both regions have a favorable geopolitical location – on a border with the EU. However, they developed in different directions regarding the process of cross-border regional cooperation. Therefore, they present interesting cases for the analysis to find out what factors accounted for the fast integration of one region, the city of St. Petersburg, into European space and resistant attitude of another region, Leningrad oblast, towards closer cooperation with neighboring European cities and countries.

St. Petersburg is a city located in northwestern Russia on the delta of the river Niva at the east end of the Gulf of Finland on the Baltic Sea. It was formerly known as Leningrad (*Ленинград*, 1924-1991) and Petrograd (*Петроград*, 1914-1924).

Today St. Petersburg is Russia's second-largest city, Europe's fourth largest city, a major European cultural center, and the most important Russian port on the Baltic. St. Petersburg is the northernmost city with over one million people. It is the administrative center of the Leningrad oblast (while being a separate region).

St. Petersburg is a seaport, which drains into the easternmost part of the Gulf of Finland (part of the Baltic Sea). It is included in the North-Western Economic Area and North-Western Federal Okrug. The city's territory, including a total of 42 islands in the Neva delta, occupies an area of 570 km² (making it the smallest of the Russia's federal components), of which waterways comprise around 10%. The population of the city was estimated 4.70 m. at 1 January 1999.

Leningrad oblast (in Russian: Ленинградская область, Leningradskaya oblast) is a constituent unit of the territorial-administrative division of Russia. It is also located in the Northwestern federal district and named after the revolutionary Vladimir Lenin. Territorially, it contains St. Petersburg but Leningrad oblast is administratively separate from it. The oblast has an area of 85,900 km² and a population 1,669,205 as of 2002 (all-Russian Population Census). Leningrad oblast managed to retain its Soviet name after the fall of the USSR.

The Leningrad oblast is also situated in the north-west of the Eastern European Plain and also lies at the Gulf of Finland. It forms part of North-Western Economic Area and the North-Western Federal Okrug. It borders Estonia and Finland. There is an international border with Estonia to the west and with Finland to the north-west and one also finds a partial border with Karelia. The oblast occupies 84,500 km² and is divided into 17 administrative districts and 29 cities. Its total population is 1,673,700, of whom 66.0 % inhabited urban areas (1999).

Map 1: Russia and Its Regions: Leningrad Oblast

Map 2: Leningrad Oblast and St. Petersburg[17]

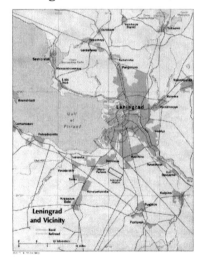

History and Politics

St. Petersburg has occupied a special place in Russian history for more than three centuries. The city was founded by Tsar Peter I ("the Great") in 1703, as a "window on the West", and it was the Russian capital from 1712 to 1918. Following the fall of the Tsar and the rise of the Bolshevik Revolution in 1917, the Russian capital was moved back

---

[17] On the Map, the city of St. Petersburg is marked as "Leningrad" (keeping the soviet name of this region).

to Moscow and in 1924 St. Petersburg was renamed Leningrad. In June 1991 the citizens of Leningrad voted to restore the old name of St. Petersburg and their decision was effected in October 1991.

On 24 April 1996 the liberal mayor of the city, Anatolii Sobchak, approved a draft treaty on delimitation of powers between St. Petersburg and the federal Government. Sobchak was defeated in the mayoral election held in May by another liberal, Vladimir Yakovlev who was reelected in May 2000, obtaining around 73% of the votes cast.[18]

Leningrad oblast was formed on 1 August 1927 out of the territories of five regions (Cherepovetskoi, Leningrad, Murmansk, Novgorod and Pskov). The region did not change its name when the city Leningrad reverted to the name St. Petersburg. While the city was a strong base for reformists and supporters of the federal Government in the 1990s, the Oblast provided support to Communists.

Autonomy

In the 1990s, the level of autonomy of each region was defined according to the RF Constitution and bilateral contracts which were signed by 50% of all regions.[19] The Constitution delineates three groups of regions according to three levels of autonomy: republics with the highest autonomy; the regions which are territorially incorporated into other regions and, therefore, subordinated to both the central government and to the administration of the "parent" region – regions with the lowest autonomy; and regions with "intermediate" autonomy – all those which do not belong in either the first or second group. Both St. Petersburg and Leningrad oblast belong to the third group. They do not have the "privileged" status of republics and they are not subordinated to the administration of another region. Both regions have the same contractual autonomy, as both of them have signed power-sharing agreements with the central government.[20]

The document defining the political structure of St. Petersburg is the Charter passed by the Legislative Assembly in 1998. The RF Constitution of 1993 extended the federalism principle to the sphere of international relations and created real opportunities for the regions to enhance

---

[18] For electoral data and politics of the regions of Russia in the 1990s, see *The Territories of the Russian Federation.* 2002. Europa Publications. Tailor & Francis Group. 3rd ed.

[19] See for example Kahn, J. 2000. *Federalism, Democratization, and the Rule of Law*; Ross. 2000. *Federalism and Democratization in Russia.*

[20] *The Territories of the Russian Federation.* 2002. Europa Publications. Tailor & Francis Group. 3rd ed.

their activities at the international level.[21] A "Treaty on the limitation of powers between authorities of the RF and the city of federal status, St. Petersburg" was signed on 13 June 1996. Article 16 covers the delimitation of powers in international activities between St. Petersburg and the federal centre:

> St. Petersburg has a right to establish international and external economic links on its own initiative or on the request of federal authorities of the RF..., has a right to conclude respective treaties (agreements) with regions of foreign federal states, administrative units of foreign states, and ministries and departments of foreign states.

In 1996 the administration of the oblast also signed an agreement with the federal Government on power delimitation between the federal and regional governments. The gubernatorial election of 1996 was won by an independent candidate, Vladimir Gustov. In September 1999, Valerii Serdyukov was elected governor of the oblast.

Cross-border Cooperation with Europe

St. Petersburg has developed most of its trading links with the Baltic republics, Finland, and other states in northern Europe. By the end of the 1990s, numerous shuttle traders hustled between the city and Helsinki, importing a wide variety of foreign products. Also, St. Petersburg had $234 million in trade with Latvia in 1997, including $160 million in exports.[22]

In 1991, the city signed horizontal agreements with other subnational governments and treaties with the organs of central governmental power of foreign states, for example with the Estonian Ministry of Trade. Now St. Petersburg has more than 50 twin-cities and partner-regions abroad. It is a twin-city of Rotterdam, Gdansk and Le Havre,

---

[21] The data on the main legal documents of the RF are derived from the *Russian Legal Texts* compiled and edited by W. Butler and J. Henderson (1998). This volume includes the full texts of such important for the current research documents as the RF Constitution; Treaty on Delimiting Subjects of Jurisdiction and Powers Between Federal Agencies of State Power of the RF and the Agencies of Power of the Sovereign Republics within the RF (31 March 1992); Treaty on Delimiting Subjects of Jurisdiction and Powers Between Federal Agencies of State Power of the RF and the Agencies of Power of the Territories, Regions, and Cities (31 March 1992); Treaty on Delimiting Subjects of Jurisdiction and Powers Between Federal Agencies of State Power of the RF and the Agencies of Power of the Autonomous Regions and Autonomous National Areas (31 March 1992); Treaty of the RF and the Republic of Tatarstan "On Delimiting the Subjects of Jurisdiction and Mutual Delegation of Powers Between Agencies of State Power of the RF and Tatarstan" (15 February, 1994); etc.

[22] *The Territories of the Russian Federation.* 2002. Europa Publications. Tailor & Francis Group. 3rd ed.

which have been of great help in integrating Russian partners into trans-regional networks and multilateral programmes. The city is a full-fledged participant in the "Baltic Troika" that involves two state capitals, Helsinki and Stockholm. Despite its municipal status, the local government organization is able to affect policy-decisions towards Russia, from within the country. The city's Legislative Assembly has brainstorming-like relations with Bavaria and Lombardy, and City Hall is interested in exchanging views with other federated units which tend to challenge their centre (e.g. Gothenburg, Baden-Wurttemberg, Schleswig-Holstein and Flanders).[23]

St. Petersburg administrative experts managed to establish many international links with European states (both neighbors and non-neighbors). St. Petersburg became a place for high-ranking international meetings, replacing Moscow. Its position in the Russian Federation has been affected by the development of the EU's "Northern Dimension" programme and other projects.

The city's administration experts managed to establish many international links after the fall of the iron Curtain. St. Petersburg's elite did not influence the process of foreign policy-making at the federal level, although Mayor Sobchak delivered speeches on international questions, presenting his own vision which differed from the Ministry of Foreign Affairs' official position.

Moscow's monopoly on prestigious international meetings ended and some of these meetings started taking place in St. Petersburg. For example, in April 2001, the Russian-German summit was held there. Vladimir Putin and Gerhard Schroeder opened the "Petersburg Dialogue", which is to become a forum for discussions among politicians, businessmen, scientists, journalists, and artists from both countries. Previously, such meetings had taken place only in Moscow.

St. Petersburg's political elite have always been oriented to democratic reform and a market economy. The city's leader – both its first mayor Anatolij Sobchak and his successor, Governor Vladimir Yakovlev – focused on radical market reform and did achieve some positive results. From the 1960s until the beginning of 1980s, the city was one of the major centers of the military industry. In the 1990s, priorities completely changed. Mayor Sobchak aimed at transforming St. Petersburg into a banking center and cultural capital and sought to provide all possible assistance to tourism, the transport sector and the sciences. The military share of industrial production was drastically reduced, with the

---

[23]   Marin, A. 2002. "The International Dimension of Regionalism – St. Petersburg's 'Para-Diplomacy'", in Kivinen, M. and Pynnoniemi. eds. *Beyond the Garden Ring. Dimensions of Russian Regionalism*. Helsinki: Kikimora Publications, p. 147-174.

exception of shipbuilding, which received large orders from the Indian and Chinese Navy.

Foreign investments have played an important role in the development of market relations. Politically St. Petersburg supported the most radical advocates of democratic and market reform during the period of the late 1980s to the first half of the 1990s. The relatively-higher cultural and educational level of the citizens played an important role.

Many concurrent changes were taking place: the rise in St. Petersburg's economic independence sub-regional integration with the institutionalization of bilateral, intergovernmental, cross-border and translocal cooperation; the development of transnational actors and translocal cooperation. All these factors made the region a significant actor in the European arena and even raised "idealistic" projects of the region's membership in the EU.

Like St. Petersburg, Leningrad oblast shares a border with the EU in the Northwest of Russia. Geographically, it is a bigger region and has a much longer border with the EU than does St. Petersburg. However, bordering Estonia and Finland just like St. Petersburg, Leningrad oblast has developed almost no links with Western partners. There have been no incentives on the part of regional political administrations to establish, for example, cultural and academic exchanges, or trade links with neighboring regions. This is even more surprising if we recall that Leningrad has a much longer border with the EU and, thus, inevitably falls in the area of the EU's Northern Dimension programme. Nonetheless, the regional administration tends to strengthen connections with the federal government than with European neighbors, and relies more on federal subsidies. The main foreign question in the region is the construction of ports which could redirect Russian trade from the Baltic countries in the future. However, this is a long-term project.[24]

There are also some features that distinguish Leningrad oblast from other regions. The location on an EU-border probably makes the region more important in terms of federal budget distribution and subsidization. The region successfully lobbied the project from the establishment of a pipeline system and ports in Leningrad oblast. In June 1997, the President's Decree announced support for the building of three sea-ports in this region (Primorsk, Ust'Luga, Barareynaya Bay).[25] Thus, it is

---

[24] *The Territories of the Russian Federation.* 2002. Europa Publications. Tailor & Francis Group. 3rd ed.

[25] Khudoley, K. 2002. "Russian-Baltic Relations – a View from Saint Petersburg", in Hubel, H., ed. *EU Enlargement and Beyond: The Baltic States and Russia.* Berlin: Verlag Arno Spitz, p. 337.

apparent that the local government tends to rely much more on the federal government than on cooperation with European neighbors.

In contrast to St. Petersburg, Leningrad oblast was slow in the implementation of market-oriented reforms. The reason for this is that Leningrad oblast is a more rural region where the absence of private land became an obstacle to economic transition. The attitude of the regional government was more resistant to reform (e.g. they have chosen to keep the old name of the region, "Leningrad oblast", instead of renaming it "St. Petersburg oblast").

Why are some of the CUs (regions) active in paradiplomacy (like St. Petersburg) and some (like Leningrad oblast) are not? Three sets of motivation are often distinguished: economic, cultural and political.[26] Economically, pro-Europe regions are interested in investment, markets for their products, technology for modernization, and in the development of tourism infrastructure. Regions with a predominant ethnic minority, with their own culture and language, seek support in the international arena. By acting more or less independently, they establish themselves as individual actors with their own identities. Finally, there might be political reasons for entering the international arena. "Those with nationalist aspirations seek recognition and legitimacy as something more than mere regions".[27]

Thus, among the potential factors which might have influenced the intentions of the regions to develop cross-border cooperation with their European neighbors, we have distinguished contextual and institutional factors. An "opportunity structure" can be constructed, based on three contextual factors: (1) the type and the level of economic development; (2) ethnicity (distinct language and culture); (3) geopolitical position; and on two domestic-institutional factors: (1) constitutional autonomy and (2) contractual autonomy. Table 1 shows the factors that might provide some explanations for the initiative of a region to develop cross-border cooperation for the two case regions, in terms of the motivation and the level of their relationships with the EU.

---

[26] Keating, M. 1995. "Europeanism and Regionalism", in Jones, B. and Keating, M. eds. *The European Union and the Regions*. Oxford: Clarendon Press, p. 1-22.
[27] *Ibid.*

Table 1. Opportunity and Motivation Structure[28]

	Sank Petersburg	Leningrad oblast
Geopolitics	North-West part of Russia Direct Border with the EU	North-West part of Russia Direct Border with the EU
Ethnicity – Ethnic composition	Russians 89.1% Ukrainians 3% Jews 2.1% Belorussians 1.9%	Russians 90.9% Ukrainians 3% Belorussians 2%
– Languages	Only Russian	Only Russian
Economics – Urban/Rural	Urban population (as of 1 Jan. 1998): 100% (Russia overall: 73.1%)	Urban population (as of 1 Jan. 1998): 65.9% (Russia overall: 73.1%)
– Rank of economic development[29]	EBRD Rank N 2	EBRD Rank N 32
Constitutional Autonomy	Intermediate level of autonomy	Intermediate level of autonomy
Contractual Autonomy	Bilateral contract with central government has been signed	Bilateral contract with central government has been signed

The geopolitical locations of both regions are the same. To a certain extent, it could be thought that Leningrad oblast has a more favorable location, as it has a longer border with the EU and, therefore, the inter-action line is longer. It also puts it in the sphere of interests of the EU and in the domain of the Northern Dimension. As for ethnicity, the two regions look pretty much the same – Russians compose majority both in St. Petersburg and Leningrad (89.1% and 90.9% accordingly) – and the only spoken language is Russian. However, economic indicators shed more light as potential explanatory variables, as this is the only indicator

---

[28] These data was published in *The Territories of the Russian Federation*. 2002. Europa Publications: Taylor & Francis Group. 3rd ed., and in Orttung, Robert W., eds. (2000). *The Republics and Regions of the Russian Federation. A guide to Politics, Policies, and Leaders.* East-West Institute.

[29] Rank of economic development is measured by the EBRD in 1998-2000. It starts from "1" for the most developed to "89" as the least developed. The group of EBRD experts elaborated a complicated system for the evaluation of the economic development of the regions encompassing not only such well-known indicators such as Per Capita Gross Regional Product but also registered unemployment (as a percentage of the economically active population), ratio of average income to poverty line, population with income below poverty line, the employment in small firms as a proportion of general employment. The EBRD. Strategy on Russia, Annex 4: Regional Indicators for 2000 and 2001 http://www.ebrd.com/about/strategy/country/russia/annex. pdf, accessed in March 2004.

which significantly differentiates the two regions. St. Petersburg is a more industrial region. However, the difference of 34% in urban population between the two regions is a large one, but it is not striking. As for institutional asymmetry, both regions have had the same level of autonomy outlined in the Constitution and both regions have signed bilateral power-sharing contract with the central government.

The following table demonstrates varieties of cross-border cooperation developed between some of the regions of Russia and European actors.

Table 2. Paradiplomatic Activities[30]

Opportunities of Paradiplomatic Activities by 1999	St. Petersburg	Leningrad oblast
Euroregions	Participating	-
Twin-cities	Participating	-
European Organizations	A member of a few organizations	-
Common projects	Numerous	-
Common programmes	Numerous	-
Trade with Europe	Developed	-
Foreign Investment (1997)	2.5% of Russian total	1.29% of Russian total
Joint ventures (1997)	9.96% (1,467) of Russian total	0.9% (134) of Russian total

This data demonstrates that St. Petersburg was actively participating in different organizations, and international programmes. In the Euroregions (for example in Baltic Euroregion), it is a twin-city with a number of European partners; it is an active member of a few organizations (Baltic Troika, the Council of Baltic Sea States, and others); it has developed trade with the Baltic republics, Finland, and other states in Northern Europe; it has launched numerous academic and expert exchange programmes; and it is involved in a number of common projects. The foreign investment in St. Petersburg is twice as high as it is in Leningrad. Finally, joint ventures in the city constitute almost 10% of all joint ventures in Russia, when the rate is only 0.9% in the Leningrad oblast.

Given that analysis includes two regions of the same country, with the same geopolitical location, this comparative study controls for contextual opportunity structures. Politically, both regions are the

---

[30] These data were published in *The Territories of the Russian Federation.* 2002. Europa Publications. Tailor & Francis Group. 3rd ed., and in Orttung, R. W. eds. 2000. *The Republics and Regions of the Russian Federation. A guide to Politics, Policies, and Leaders.* East-West Institute.

subjects of the same legal and administrative system. Culturally, they present the same ethnic groups – Russians are the predominant group within both regions. There is not doubt that ethnic reconfirmation was never an issue on the political agenda of the regional administrations.

The economic, geopolitical, cultural, legal, and functional contexts are the same for both regions. The important question, then, is: what other factors could determine the different strategies of cross-border cooperation with their European neighbors?

To sum up, we can make the following conclusions regarding the factors "encouraging" cross-border cooperation with Europe. The factors that seem to be favorable to the development of cross-regional cooperation and communication are political orientation of the regional government and the predominance of rural or urban types of production. Smaller size might also make it easier to adapt to the requirements of transition to the external, European environment. However, such factors as location in the European part of the RF, and even on its northwest borders with the EU, are, apparently, not sufficient (pre-)conditions for the development of successful foreign policy towards European neighbors. Having additional institutional autonomy (through bilateral treaties signed with the central government in Moscow) is a necessary but not a sufficient condition either.

Thus, the questions posed in the next section are: What is the impact of cross-border cooperation and communication with European partners on regime transition within the regions? What have the regions gained from cooperation with the EU in terms of policy learning, adaptation of democratic values, and democratic administration? What was the impact of close cooperation with European actors during the 1990s? To answer these questions, we analyze the success of democratization, by 2000. In other words, we attempt to measure the degree of democratization as an impact of Europeanization measured by compliance with the norms and values of European acts. We have chosen the freedom of the mass media as one of the most important parameters of democratization and the compliance with laws and practices adopted by the regional administrations with The *European Convention on Human Rights*.

IV. The Impact of Cross-border Cooperation on Sub-national Democratization: Freedom of Mass Media as a Case-study

Democracy includes many parameters – division of powers, fair elections, party competition, regular executive turnover, independent mass media, freedom of speech, etc. However, in this study we do not deal with the established democratic context but with the so-called country in transition. Therefore, we address "democracy" and the necessary prerequisites of consolidated democratic regime within the manageable framework of the freedom of mass media in the regions. In the regions of Russia during the 1990s, local "rules of the game" often had nothing to do with the federal legislation. Thus, the role of the media did not correspond to the niche defined by federal law and reflects the regime transition within the region. This arena allows us to focus on the "domestic" peculiarities of the specific regional case regimes.

A truly democratic regime in a federal state would apply democracy on the local, i.e. regional, level which would be associated with mass media freedom.[31] *The European Convention on Human Rights* defines freedom of speech as "The right to free expression, the right to receive and disseminate information and ideas". The mass media has the main function of being a mediator between the society and the authorities and it is a catalyst for change in society. Therefore, the freedom of the mass media in the regions seems to be quite objective criteria for measuring the regime within those regions. The data for the regions are accumulated by the Public Examination Global Project (PEGP). The Survey was conducted by the Russian Union of Journalists, the Glasnost Defence Foundation, the National Institute for Socio-psychological Studies, The Mass Media Law and Policy Centre and ANO Internews. The project staff created an index measuring freedom of mass media for each region that incorporated a few indexes: freedom of access to information, freedom of production of information, share of independent mass media in a region, etc. The maximum score for each measurement – freedom of access, freedom of production and distribution of information – was 100.[32] The following table demonstrates the degree to which freedom of speech varies across the regions.

---

[31] Following the tradition of the experts of PEGP, this study uses the notions of "freedom of speech" (FoS) and "freedom of mass media" (FMM) interchangeably.

[32] Data on Mass Media was accessed on 9 May 2004 at http://www.freepress.ru/ win/public/6htm. These data (index of press freedom across the regions) were archived by the Helsinki group, a summary can be found at: http://www. freepress.ru/arh.shtml Accessed on 1 May 2006 (Анатомия свободы слова за 1999-2000 год).

Table 3. Index of the Freedom of the Speech:
Regions with the Highest and Lowest Scores

Region	Freedom of Access to Information	Freedom of Production of information	Total Score
Moscow	23.0	75.0	62.9
St. Petersburg	60.0	56.5	50.2
Vladimir oblast	66.0	63.2	49.3
Vologda oblast	57.5	56.9	49.1
Sverdlovsk oblast	57.0	51.6	47.3
Agin-Buryat AO	14.0	-	19.5
Mari-El Republic	20.0	28.7	19.3
Magadan oblast	18.0	28.8	19.1
Chukotka AO	25.0	-	18.8
Karachaevo-Cherkessia Republic	21.0	2.3	14.6

Table 4. Independent TV and Radio and Total Freedom
of the Production of Information. Regions
with the Highest and Lowest Scores

Region	Independent TV(in % from total in the region)	Independent Radio(in % from total in the region)	Total Freedom of Production of Information
Moscow	67.0	96.0	75.0
Vladimir oblast	22.0	43.0	63.2
St. Petersburg	67.0	25.0	65.5
Republic of Adygeya	0	0.6	7.5
Republic of Karachaevo-Cherkessia	2.8	1.7	2.3
Yamalo-Nenets Autonomous okrug	0	0	-

The following table summarises the descriptive statistics for Freedom of Speech in the regions.

Table 5. Data Summary: Freedom of Speech,
Independent TV, Independent Radio

	N of Observations	Min	Max
Freedom of Speech	87	15.00	63.00
Independent TV (in % from total in the region)	87	0.00	67.00
Independent Radio (in % from total in the region)	87	0.00	96.00

The results of the PEGP project demonstrated wide disparities across the regions of Russia, with the most democratic being St. Petersburg and Moscow (with total score of the freedom of speech 50.1 and 62.9) and the least democratic Chukotka and Karachaevo-Cherkessia (18.8 and

14.6 respectively). The result of the project demonstrated that there was no region that would meet all criteria of the freedom of speech. However, two regions have favorable conditions for access to information as required by the press (Yaroslavl and Murmansk Regions), and one region has a favorable climate for both the production and dissemination of information (Moscow). The data collected in this project, seems to be quite an objective indicator to use in further speculation on the regime change in the regions.

Table 6 demonstrates some of the parameters measuring the freedom of mass media across the two regions of Sank-Petersburg and the Leningrad oblast. Here, we also have the chance to analyze how compatible the two regions are with the European Convention.

Table 6. Measuring Freedom of the Mass Media across two Regions

Region	Freedom of access to information	Freedom of production of information	Freedom of the press (assessment based on regional laws)
St. Petersburg	60%	57%	50%
Leningrad oblast	58%	34%	42%

The freedom of the mass media includes freedom of access to the information and freedom of production. The last column summarizes the final score of the freedom of mass media. The disparities between the two regions are quite significant. It seems the regional administration of St. Petersburg had adapted their regional legislation regulating mass media activities to the democratic standards by 2000, while Leningrad seems to score much lower than its neighbor.

During sub-national regime transition, regional authorities often try to adapt liberal federal legislation to their own needs, issuing repressive local orders that restrict freedom of mass media and infringe the right to freedom of speech. This parameter was also incorporated into the system of indexes. It is especially valuable information because it includes the thorough analysis of the local laws.[33]

The right to seek and receive information freely is dependent on unrestricted access to information, transparency of executive, representative and judicial authorities, the response of officials to requests for the information, and the fairness of accreditation requirements. The Survey created an index of the level of *free access* to the information in each region of the RF by analyzing local laws and the practices involved in facilitating access to information. The difference between the two case

---

[33] *Ibid.*

areas, St. Petersburg and Leningrad is not necessarily a striking difference but it is important to underline. There is, in fact, greater access to information in St. Petersburg.

Second, the index of the *production* of information is measured by the analysis of the regional registration regulations (broadcast licenses), local tax and other codes which affect the media's economic activities, and the government's role in regulating access to the means of production, both print and electronic. Here, there is an evident gap of 23 points between the two regions. This indicator may be a more objective criteria for measuring press freedoms as this parameter was based on the analysis of:

– regional laws regulating media activities,

– the analysis of regional accreditation rules for journalists,

– field research in regional markets (experts collected data on number, circulation and ownership structures of print media; number, capacity, ownership structures of publishing houses; number, signal capacity, coverage area and ownership structures of TV and radio broadcasting companies; and information on the terms and conditions for granting state support to mass media in each market). The environment established by local administrations for distributors of the press (includes analysis of tax system and other privileges for distribution and the number of permits needed to open press outlets).

Another criteria of pro-democratic tendencies in the regions is the percentage of the independent mass media as a proportion of total mass media (which are subdivided here into percentages of state-regulated and independent). The data on the percentage of the independent mass media in St. Petersburg and Leningrad is summarized in Table 7.

**Table 7. Share of Independent Mass Media
in St. Petersburg and Leningrad**

Regions	Independent TV (in % from total in the region)	Independent radio (in % from total in the region)	Independent press (in % from total in the region)
St. Petersburg	67%	25%	90 %
Leningrad oblast	0.2%	1%	54.2 %

There is a striking difference between the two regions. In Leningrad, it seems there are no independent TV channels and radio programmes at all: the proportion of independent TV is just 0.2%, and just 1% for independent radio, compared to 67% and 25% in St. Petersburg, accordingly. The gap in the percentage of the independent press is also very significant: 90% of the press in Petersburg is independent while in Leningrad it is only 54%.

The indicator is another significant criterion for judging the success of the democratization process in the two case regions. The role of the independent mass media increases over a pre-election period, for example, when it has the potential to shape public opinion and to influence the result of the election. If freedom of the mass media is suppressed, it leads to a situation in which the public is given unreliable information which might secure the victory of those in power. Thus, press, radio, and TV are turned into a tool for settling scores in political battles, without real care about everyday needs of the general public. In this regard, the citizens of St. Petersburg seem to be in a more beneficial situation than those living in Leningrad oblast. Consequently, not only has the former case followed a more internationally-oriented agenda, but it also is rated higher on traditional indices for democratic stability.

V. Conclusion

Both St. Petersburg and Leningrad oblast are located in the northwest of Russia and both share a border with the EU. Both have signed bilateral power-sharing agreements with the central Government in Moscow and, therefore, both were granted some additional autonomy in managing their domestic and foreign policies. Then, while having the same favorable geopolitical location and the same status and autonomy within the federation, why have they developed such disparate policies towards Europe?

It is important to note that during the 1990s there were some discussions of a merger of St. Petersburg and Leningrad oblast, but all negotiations reached a deadlock. The reason lies with the divergent ambitions of municipal and regional officials since both CUs have chosen different models of social-economic reforms which have reached different results.[34]

The territorially bigger Leningrad oblast has a longer border with the EU. Nonetheless, it failed to achieve the same attention and "popularity" in cooperation with Europe as St. Petersburg. There are some important explanatory factors. First, St. Petersburg has greater economic development and more industry while the oblast is mainly a rural region. Second, these two cases demonstrated the important and probably decisive role of the political orientation of the local elites – as the

---

[34] Khudoley, K. 2002. "Russian-Baltic Relations – a View from Saint Petersburg", in Hubel, H. ed. *EU Enlargement and Beyond: The Baltic States and Russia.* Berlin: Verlag Arno Spitz, p.334.

governments of these two regions reflected different geopolitical priorities. Amazingly within the same country, even within the same geographic area, there is the coexistence, side by side, of two different types of development, and two different types of attitudes toward the European neighborhood, each with different vectors of political regime development.

This study shows that two regions sharing a number of common features may exhibit different trends of what have been once defined as "old" regionalism and "new" regionalism.[35] Leningrad oblast presents a classical example of old regionalism, as the one rooted in tradition, conservative political values, and ruralism. During the 1990s, the region was predominantly rural and kept communist traditions and values of the Soviet system, such as common agricultural lands and production in the forms of "sovkhozy" (soviet administration of agriculture) and "kolkhozy" (collective administration of agriculture). In contrast, the region of St. Petersburg is an example of the new regionalism as it is characterized by modernization, change, and urbanism. Privatization, a transition to a market economy, democracy, and openness to the establishment of contact with European neighbors are dominant characteristics of the region. Thus, the phenomenon of "new regionalism" seems to go far beyond the EU's borders, as it affects neighboring areas, even in non-candidate countries.

The two units have chosen different models of social-economic reforms and they have had different results. St. Petersburg has, in fact, a more developed economy while the Leningrad oblast suffers from an economic crisis (agricultural products cannot compete with imported products, there are low general living standards, the people's minimal purchasing capacity restrains the development of trade and social services, etc).[36] In the latter, the military industry and heavy industry clearly failed to properly convert to civilian production.

Three important conclusions can be derived from this comparative study. The first conclusion we make from the analysis is regarding the importance of geopolitical location. This variable is not a sufficient factor for successful development of cross-border cooperation. Location on the border with the EU, just as location in the north-west part of the RF, is not enough to promote effective development of cross-border cooperation. Geopolitics does not seem to matter. In contrast, the political orientation of the regional governments and the strategies of politi-

---

[35] Keating, M. 2002. "Territorial Politics and the New Regionalism", in Heywood, P., Jones, E. and Rhodes, M. *Development in West European Politics 2*. Basingstoke: Palgrave, p. 201-220.

[36] Khudoley, K. 2002, p. 334.

cal elites seem to play a crucial role in determining openness, or lack thereof, to democratic European neighbors.

The second conclusion regards the role of institutional mechanisms of subnational regionalization, namely the establishment of constitutional and contractual asymmetrical federalism. Asymmetric federalism was the institutional design which was expected to encourage the development of cross-border cooperation. Both case regions included in this study have had the same degree of constitutional and contractual autonomy. Both regions have signed bilateral power-sharing agreements (contracts) with the central government, which outlined their spheres of autonomy. Thus, it seems that granting additional institutional autonomy to a region is not a sufficient condition to initiate cooperation with neighboring regions and countries. On the other hand, if this additional institutional autonomy in the form of a contract had not been granted to St. Petersburg, it would not have been able to develop such a successful cooperation with European partners at all. Thus, although additional autonomy is not a sufficient explanatory variable on its own, it is, no doubt, an important factor in the development of cross-border regional cooperation.

Third, there are a few different forms of regionalization that have taken place in the transition period of the 1990s. The first one is the regionalization *within* the Russian Federation and initiated by the central government, through the establishment of constitutional and contractual asymmetry (subnational regionalization). Then, there is the regionalization within the RF, but initiated by the EU and European countries (Finland) through the introduction of the Northern Dimension (through cooperation of the regions of the RF with regions of neighbor-states – cross-national regionalization). Both national and subnational regionalization have encouraged the development of cross-national regionalization.

The first form of regionalization was characterized by decentralizing reforms and the establishment of asymmetrical federalism. The second form of regionalization can rather be defined as integration through cooperation and communication (along with Karl Deutsch theory of integration). Thus, we have analyzed two forms of regionalization: decentralization within the state and the formation of cross-border regions, (through cooperation and communication). The first has certainly impacted the second form of regionalization. Decentralization, as a form of democratization of center-periphery relations within the federation, in fact, accompanied cross-border regional cooperation and the conversion of many regions (such as St. Petersburg, Karelia, Pskov, and other regions of the RF) into Euroregions.

The theoretical conclusion, then, is that cross-border cooperation, analyzed in the context of "Europeanization" and "democratization", seems to present "the-chicken-and-egg-dilemma": the precondition to the start of cross-border cooperation with Europe is pro-democratic orientations of regional administrations and the outcome of such communication is further adaptation to democratic values. An actor in such cooperation, a region, should already be oriented pro-democratically. However, it is through the process of interaction with Europe that a region can strengthen democratic tendencies.

Thus, the evolution towards democracy or autocracy adopted by regional elites in the 1990s had a defining effect on the development of cooperation with European actors. The case study of two regions in this analysis confirmed this assumption.

This chapter has focused on the experiences in cross-border cooperation of two northwestern Russian regions in the context of theories Europeanization (understood as the adaptation of democratic norms and values through regular cooperation with European actors) and democratization (defined as the impact of such cooperation on regime transition in the regions). It has presented some relevant findings, regarding the conditions, limitations, and regional impact of involvement in European politics through the development of cooperation. The study suggests that with regard to the democratization process in Russian regions, cross-border cooperation with Europe seems to have some positive effects that strengthen pro-democratic tendencies. They also facilitate local regional dialogues through the implementation of horizontal partnerships and the creation of sub-national organizations. Cooperation between the regions of Europe, through EU programmes, Euroregions, and specific projects, prepares the ground for further implementation of pro-democratic politics and initiates a spillover process, at the sub-national level.

References

Commission of the European Communities, "The European Union and Russia: the future relationship", COM (95) 223 final, 31 May 1995.

Cowles, Maria Green, James A. Caporasso, and Thomas Risse, eds., (2001). *Europeanization and Domestic Change.* Ithaca: Cornell UP.

Deutsch, Karl W. 1953. *Nationalism and Social Communication. An Inquiry into the Foundations of Nationality.* The M.I.T. Press.

Deutsch, Karl W. 1966. *Nationalism and Social Communication.* The M.I.T. Press, Massachusetts Institute of Technology: Cambridge, Massachusetts, and London, England.

Haukkala, Hiski, "The Making of the European Union's Common Strategy on Russia" in Haukkala, Hiski and Medvedev, Sergei, eds. *The European Com-*

mon Strategy on Russia. Learning the Grammar of the CFSP. The Finnish Institute of International Affairs, Helsinki, 2001:66

Keating, Michael. 2002. "Territorial Politics and the New Regionalism", in Heywood, Paul, Jones, Erik and Rhodes, Martin. eds. *Development in West European Politics 2.* Palgrave, UK, p. 201-220

Keating, Michael. 1995. "Europeanism and Regionalism", in Jones, Barry and Keating, Michael, *The European Union and the Regions.* Oxford: Clarendon Press, p. 1-22

Khudoley, Konstantin. 2002. "Russian-Baltic Relations – a View from Saint Petersburg", in Hubel, Helmut. ed. *EU Enlargement and Beyond: The Baltic States and Russia.* Berlin: Verlag Arno Spitz, p. 337.

Kivikari, Urpo. 1999. "The Application of growth triangle as a Means of Development for the Kaliningrad region", in Kivikari, U., Lindstrom, M. and Liuhto, L. eds. *The External Economic relations of the Kaliningrad region.* Turku: School of Economics and Business Administration, Institute for Easr-West Trade C2, p. 1-13.

Moravcsik, Andrew. 1999. *The Choice for Europe: Social Purpose and State Power from Rome to Maastricht.* Ithaca, NY: Cornell Univerity Press; and Heritier, Adrienne. *Policy Making and Diversity in Europe. Escape from Deadlock.* Cambridge: Cambridge University Press.

Marin, Anais. 2002. "The International Dimension of Regionalism – St. Petersburg's "Para-Diplomacy"", in Kivinen, M. and Pynnoniemi. eds. *Beyond the Garden Ring. Dimensions of Russian Regionalism.* Helsinki: Kikimora Publications, p. 147-174.

Morlino, Leonardo. 2002. "The Europeanisation of Southern Europe", in Costa Pinto, A. and Teixera, N.S. eds. *Southern Europe and the Making of the European Union 1945-1980.* New York: Columbia University Press, p. 237-260.

Ladrech, Robert. 1994. "Europeanization of Domestic Politics and Institutions: The Case of France". *Journal of Common Market Studies* 32 (1):69-88

Obydenkova, A. 2006, "Democratization, Europeanization and Regionalization beyond the European Union: Search for Empirical Evidence". *European Integration online Papers.* Vol. 10, No. 1.

Obydenkova, A. 2005. "Puzzles of the European Regional Integration and Cooperation: Interplay of "Internal" and "External" Factors", in Di Quirico, Roberto. ed. *Europeanisation and Democratisation. Institutional adaptation, Conditionality and Democratisation in European Union's Neighbor Countries.* European Press Academic Publishing, p. 199-222.

Obydenkova, A. 2005. "European Integration, Regionalisation, and Democratization: Investigation in Dependent and Independent Variables". *Working Paper of the European University Institute.* Florence, Italy: European University Institute, SPS 2005/06. ISSN: 1725-6755.

Obydenkova, A. 2005. "The Role of Transnational Regional Cooperation in the Regime Transition in the Regions: Regions of Russia and the EU". *Working Paper of the European University Institute.* Florence, Italy: EUI, European University Institute, SPS 2005/05. ISSN: 1725-6755.

Obydenkova, A. 2005, "Institutional tools of conflict management: Asymmetrical Federalism in Ethnic-Territorial Conflicts: Quantitative Analysis of Russian Regions". *Peace, Conflict and Development: An Interdisciplinary Journal*", Issue 7, July 2005

Obydenkova, A. 2004/16. "The Role of Asymmetrical Federalism in Ethnic-Territorial Conflicts". *Working Paper of the European University Institute, EUI*. Florence, Italy: European University Institute, SPS 2004/16.

Obydenkova, A. 2004/20. "The Paradox of Democratization and Federalization in the Russian Regions". *Working Paper of the European University Institute*. Florence, Italy: EUI European University Institute, SPS 2004/20.

Obydenkova, A. 2006. "New Regionalism and Regional Integration: The Role of National Institutions". *Cambridge Review of International Affairs*, Vol. 19, No. 4 (December).

Orttung, Robert W. ed. 2000. *The Republics and Regions of the Russian Federation. A guide to Politics, Policies, and Leaders.* East-West Institute.

Risse, Thomas. 2000. *EUI Working Paper*. Florence.

Sergounin, Alexander. 2002. "Russia's regionalization. The interplay of domestic and international factors", in Herd, Graeme P. and Aldis, Anne. *Russian Regions and Regionalism. Strength through weakness.* London and New York: Routledge Curzon, Taylor & Francis Group.

Svob-Dokic, Nada. 2005. *Europeanization and democratization: The Southern European Experience and the Perspective for New Member States of the Enlarged Europe.* Paper Contribution to the CIRES Conference "Europeanization and Democratization". Florence, Italy, 16 June – 18 June.

The Territories of the Russian Federation. 2002. Europa Publications: Taylor & Francis Group. 3rd ed.

Timmermann, CF. 1996. "Relations between the EU and Russia: The Agreement on Partnership and Co-operation". *Journal of Communist Studies and Transition Politics* 12:196-223

Treaty of the European Union (ratified 1993), Copenhagen: European Council 2002

The EBRD. Strategy on Russia, Annex 4: Regional Indicators for 2000 and 2001 http://www.ebrd.com/about/strategy/country/russia/annex.pdf, accessed 9 March 2004.

CHAPTER 3

The Need for New Water Management
Structures in North America

Carmen MAGANDA

I. The Impact of Regional Integration on the Institutional
Development of Water Resources in North America

North America is characterized by multiple regional development and integration initiatives. The North American Free Trade Agreement (NAFTA) is the most extensive geographically, as it encompasses Mexico, the United States and Canada. It is, by no means, however, the only regional integration project. In fact, throughout the continent many different cross-border institutions and programs have been developed in different policy arenas outside of NAFTA's framework. One such area where advances have been made is the field of water resources management. Since the middle of the 20th century, a regional water policy system, the basis of which is the 1944 Treaty for Utilization of Waters of the Colorado and Tijuana Rivers and of the Rio Grande, has been constructed through incremental steps. Since the signing of the 1944 treaty regional management structures have been created such as, the International Boundary and Water Commission, the Border Environment Cooperation Commission (BECC) and the North American Development Bank (NADBank). While these advances are significant, there remains much work to be done as the fluid nature of water makes it one of the most difficult resources to distribute in cross-border areas.

"Borders", "states" and "regions" as political constructs often contradict each other and create political turmoil. In the field of environmental politics this is especially true because geographic frontiers rarely match political or administrative ones. This incongruence has led to significant environmental debates regarding the impact of regional integration on natural resources because water is a fundamental resource for human development which is, unfortunately, as problematic as it is necessary. Water basins frequently fit into cross-border geographic regions that have different management agencies. Stakeholders, industries, and urban and agricultural actors attempt to appropriate the water that they deem necessary for their activities and they frequently pressure their representative institutions to negotiate the allocation of greater amounts of this resource. Because of its aforementioned fluid nature, water crosses the political and administrative boundaries created to manage its distribution and protect basin levels, thus, increasing competition and creating regional conflicts.

Border water problems are becoming unexpected "threats" to social cohesion and community development in border areas over the last decade. As stated in the introduction to this volume, political decision-making regarding the distribution of this resource rarely considers cross-border impacts. This chapter argues that attention to environmental issues, such as water management, needs to cross the limits of political borders. Environmental institutions and agencies need to shift their analysis towards "cross border water regions" as social-environmentally constructed territories.

This chapter responds to the following research question: What mechanisms characterize water management agencies in cross-border regions and what institutional changes can be made to make water allocation more equitable? Whereas water management in Europe has generally been viewed as a model of stability and cooperation (even during periods of war), North America has been characterized by significantly less sharing. In fact, national water management structures can hardly be considered democratic. Facing the challenges of urban/economic growth and limited regional water accessibility, government officials often follow short-term priorities reflecting institutional interests and political pressures. Moreover, water debates often lack accurate public information on which citizens can make informed decisions. For these reasons, water resources are often wasted and social risk situations are created amongst certain sectors of local populations.

Obviously, North American water distribution issues are directly related to regional integration. On the one hand, proponents of integration contend that regional responses are necessary to address cross-border issues, such as environmental ones. Thus, international organizations and transnational social movements have recently started campaigns to protect water as an international public good. Conversely, opponents of regional integration argue that this process facilitates the economic growth contexts that pressure officials into decisions that negatively impact the environment.

This chapter theoretically addresses the relationship between regions, borders and states in North American water politics. It argues that water distribution can best be understood through the analysis of the behavior of political elites and water officials responsible for allocation decisions. The study compares this behavior in different political and administrative frameworks.

Unlike other works that examine water conflict in single governmental arenas, this project compares elite behavior across levels of government in order to identify patterns that can be useful for understanding why water distribution is so problematic in North America. It analyzes the relationship between regions, borders, and states within an analytical framework based on the examination of political power in so-called democratic institutional structures. By doing so, I introduce the concept of *border water politics*, which refers to water management in basins that cross political boundaries that demarcate countries, states, or municipalities. Even though water agencies in shared basin contexts should be characterized by public access and inter-institutional cooperation, they are often rigid bureaucratic structures with few access points for inter-institutional exchange and public intervention. By opening this "black box", I attempt to identify mechanisms necessary for proper *border water governance*. Specifically, I examine the interaction between the nature of legal/political structures and the informal behavior of political elites and water officials.

Research Design and Methods

The research agenda behind this article addresses three inter-related research questions: 1) How do government officials distribute water in shared basins? (border water contexts) 2) How can we continually ensure adequate water supplies in regions characterized by demographic and economic growth? 3) How does sub-national governance affect water politics in the national and international arenas?

In order to respond to these questions, this study examines water politics in two basin regions related to Mexican water politics: the Lerma-Chapala Basin in central Mexico and the Colorado Basin at the United States (US)-Mexico border. Each area can be considered a critical case that provides valuable lessons for the analysis of decision-making processes concerning water distribution. By comparing these cases, I can study the processes and the behaviours of actors within different border contexts (one binational and one sub-national), thus, isolating political power as an explanatory variable. In order to collect data for this project, I conducted archival research at local universities, state and local water agencies, the morgues of local newspapers and I interviewed water officials, and representatives of both local government and non-governmental organizations active in local and regional water politics. The study is based on ethnographic research within governmental structures.

95

The cases were chosen for specific structural characteristics. First, both of these water basins are characterized by significant urban and economic growth and water shortages caused by the overexploitation of aquifers. In Lerma Chapala, five Mexican states compete for water resources from this basin. Similarly, the Colorado Basin supplies water to seven US states and the Mexicali Valley in Mexico. Second, water distribution in both basins is regulated by cross-border administrative structures that represent different political bodies (Mexican states in the Lerma Chapala Basin Council, US states in the Colorado Compact, and the US and Mexico in the Binational Water Commission) as well as local and regional water authorities (each state and city that draws water from the Lerma Chapala Basin has its own water authority whereas regional authorities, such as the Metropolitan Water District complement state and local agencies in the US). In order to understand intra-case competition, I have chosen two cities in which to study water decision-making. In the Lerma Chapala Basin, I have examined water politics in Silao, Guanajuato. This city is growing rapidly due to agricultural activity and above all, the installation of a General Motors plant, which has significantly increased the municipality's water needs. Similarly, in the Colorado Basin, the city of San Diego has witnessed a demographic and economic boom which has forced it to import between 75% and 90% of its water resources since 1947.

Finally, water politics in both cases are currently being transformed by new forms of elite behavior, which means that both basins are currently passing through a critical time period in which "the rules of the game" are being changed. Once these new "rules" are incorporated into water allocation structures, they will dictate how water decision-making occurs for years to come. In the case of Mexico, recent administrative changes have led to state challenges to the traditional authority held by the central government in water matters. Similarly, regional integration in North America has weakened institutional constraints on local officials creating more freedom for self-interested behavior and increased competition (see table one below) Thus, the creation of instruments for democratic, cross-border water governance is not just a theoretical exercise, but an immediate policy necessity. This chapter argues that new democratic structures, that cross administrative and political boundaries, need to be created to better reflect basin needs and ensure the long-term protection of water resources.

Table 1. Research Design

Constant Variables	Guanajuato-Lerma Chapala Basin	San Diego-IID Colorado River Delta
Demographic and socio-economic growth	Industrial growth increased population of Guanajuato State from 1,000,000 in 1940 to 4,600,000 in 2000	Technological industries in San Diego County increased population from 300,000 in 1950 to 3,000,000 in 2005
Basin Users	Basin shared by five states	Basin shared by seven US States and Mexico
Openness of water debates	No traditional debate-just beginning	Limited debate in periods of drought
Structural Framework Varied		
Institutional Flexibility	Traditionally Centralized	Federal System Binational, National, Sub-national actors
Institutional Stability	Democratization process led to state challenges (Fox)	Local conflict created by unclear binational accords

II. Water Politics in North America: Bridging Two Bodies of Literature

The literature on border water politics in North America is growing and dynamic. However, like the process of regional integration in this field itself, it cannot be considered complete. Many interesting individual studies have been published. However, when viewed globally, the literature in this field remains disjointed with little dialogue between the supranational and sub-national arenas.

Many works from varied sources focus on the impact of regional integration on the management of water resources at the national and supranational levels. These studies include academic research, reports from policy organizations, collegiate sources and publications from environmental centers, such as the Water Governance program (part of the UNDP: United Nations Development Programme), the Global International Waters Assessment (GIWA), the International Water Resources Association (IWRA), the Utton Transboundary Resources Center, the Pacific Institute and the the North American Congress on Latin America (NACLA), to name a few. Within this world of trans-boundary water management information, there is a lack of specific analysis on water politics in North America as an integrated region. Instead, the literature is characterized by a focus on distinct geographic boundaries (Canada-US or US-Mexico) and a lack of integration in management of hydric resources in these specific border areas.

Research on the US-Mexico border also demonstrates another division within the literature as materials produced on different sides of the border are quite distinct. Most analysis of binational water management coming from scholars on the US side focuses on the institutional agreements that regulate water distribution between the United States and Mexico. One finds valuable materials on this topic in many of Stephen Mumme's works (2002, 2003, 2005), where he analyzes the main concerns regarding the binational water agreement as well as the institutional and local capacities for transboundary water management and environmental policy-making. Also, another political scientist who focuses on water issues, Helen Ingram (1990, 1995 *et al.*, and 2001 with Battler Joachim), has produced interesting reflections on the contemporary challenges to modern governance in transboundary waters, basically along the US-Mexican border. Her work argues that there is a need to develop a more integrated and participatory approach to managing binational water resources. Similarly, other US authors, such as Christopher Brow (2003), Rick Van Shoik (2004), Suzanne Michel (2003), Mark J. Spalding (1999) and Robert G. Varady (1998) examine institutional development and political interest representation in binational water resource management. In general, these works utilize a "top down" approach as they examine binational water agreements and their (lack of) implementation.

Conversely, the literature on border water management that is presently being produced in Mexico takes a decidedly "bottom up" approach. Recent works generally focus on local actors and local practices, and they do not necessarily address the institutional arena. For example, works by Aboites (1988, 1994, 1995 and 1998) and Walsh (2005) describe problems related to irrigation and population in border areas. The theme of water as a public service, marginalization and social conflict, has been developed by Vivienne Bennett (1995, 2002), Oscar Pombo (1999, 2002) and Elizabeth Mendez (1993). Only a few authors, such as Stephen Mumme (1993), Linda Fernandez and Richard T. Carson (editors 2002), I. Aguilar Barajas and S. Mumme (2003), and Suzanne Michel (editor 2002) incorporate analysis of binational water management. From a theoretical point of view, this article attempts to combine these two approaches into an integrated study of the institutional arrangements concerning binational water management and the behavior of local political actors within these structural contexts.

III. Theoretical Framework: Complexity, Competition and Cross-border Comparisons in North American Water Politics

As stated above, current scholarship on water politics in North America is characterized by a distinct boundary. In the case of the US-Mexico border, most analysis of binational water management focuses on the institutional agreements that regulate water distribution between the United States and Mexico. These studies accurately discuss national competition for water resources and the development and impact of institutions at the border (see Mumme, Ingram and Garcia Acevedo, Boime, Spalding). Moreover, the official reports available to researchers focusing on the asymmetries of US-Mexico institutional arrangements examine the management of binational water resources without addressing the behavior of political actors. For example, the material produced by the Commission for Environmental Cooperation (CEC),[1] particularly the *North American Boundary and Transboundary Inland Water Management Report* (Law and Policy Series # 7), has clearly established the different institutional frameworks for border water politics in each participant country. In the Mexican case, water management is a centralized fact regulated by the National Water Commission (CONAGUA), despite some efforts to decentralize water management through the 1992 National Water Law. On the contrary, water allocation in the United States is basically a matter of state legislation where surface water is generally managed separately from groundwater. This represents a constant problem due the strong relationship between these two ecosystems with separate administrations.

Conversely, domestic studies of water politics, especially those focusing on Mexico, predominantly address public competition for water resources amongst local actors outside of political institutions. Some scholars, such as Anton (2003), Davis (2002), Dobkowski (2002), Simioni (1999), Hamilton (1993), etc. investigate the creation of social or environmental risk caused by uncontrolled economic growth in urban areas. Others, such as Wescoat *et al.* (2002), Berry (1997), Ingram (1990), Brown, Lee and Ingram (1987), focus on social marginalization amongst specific sectors of local populations, most notably ethnic groups, and the injustice of domestic water politics.

Thus far, no literature has developed that addresses these two arenas of water politics within a comparative framework. This chapter argues that the notion of borders in the study of water politics needs to be expanded normatively. Rather than focusing exclusively on national borders, I argue that administrative boundaries of all types create elite competition which is the basis of water decision-making. Thus, I introduce a broader concept of "border water politics" defined as the competition for water resources between political actors in areas that are divided by administrative boundaries.

---

[1] CEC is an international organization created by Canada, Mexico and the United States under the North American Agreement on Environmental Cooperation (NAAEC).

By comparing competition for water resources in binational and sub-national contexts, this chapter examines the mechanisms that dictate decision-making concerning water politics in different political arenas. The analysis presented below demonstrates that the behavior of political elites follows similar patterns in both arenas.

Thus, "border water governance" refers to the mechanisms of decision-making in border regions, both national and sub-national, which are identified through the examination of formal and informal limits on the freedom of action held by policymakers. This chapter contends that competition occurs at the local level in border areas and it significantly influences water politics in the national and supranational arenas where laws and institutions are created to regulate what is essentially, sub-national conflict. In response to the "top down" studies normally carried out in the areas, I argue that a "bottom up" approach is more effective. The empirical evidence presented below demonstrates that water management in shared basins is characterized by two traits: a) official political behavior where elites follow self-interested strategies and b) non-institutional solutions to daily water challenges (the "neighborhood" context) that reflect informal practices.

This theoretical approach incorporates views from two schools of comparative politics. First, it borrows a major paradigm from the literature on political economy. Authors in this field, such as Max Weber, Douglass North and Margaret Levi correctly argue that even though institutions were created to limit the behavior of political actors within the framework of policy competition, these bodies are, in reality actors themselves. Hence, the leaders who make decisions within regulatory institutions actually follow incentives that rarely reflect the common good. In the case of shared basins, "common good" refers to sensible sharing of limited water supplies. Instead, the actors present in basin management structures usually attempt to maximize the amounts of water allocated to their own constituencies.

In contrast, "neighbourhood" policies discuss why water sharing occurs within competitive frameworks when rationally, it should not. This inquiry relates to shifts in the notion of political power. According to classical definitions of this term, power is "the ability to obtain your desired objectives despite resistance". Of course, not all power is based on coercion. Joseph Nye's study of "soft power" correctly examines "the ability to shape the preferences of others" as an important element of political competition. Thus, local actors often create water policies through daily practice, even though legal agreements are lacking in specific arenas.

Through the study of water management in shared basins, both in the sub-national and binational arenas, this chapter demonstrates that the conflicts that limited water supplies have generated are essentially local in nature. For many years, "unofficial" solutions were found through informal cooperation at the sub-national level. However, incomplete institutional reform and asymmetrical power relationships have fundamentally changed the framework within which water distribution occurs. In the Colorado River Basin, urban growth (especially in San Diego County) has increased water needs and recent decisions reflect the favourable position of US border cities in US-Mexico water relations (and US-Mexico politics, in general). Similarly, agro-industries, industrial parks and urban growth have increased pressure on water supplies from the Lerma Chapala Basin. Due to state challenges to the previously centralized water management system, sub-national leaders have increased their influence in the allocation system for water from the basin. Unless structural limits are created in both arenas, sub-national political elites will follow self-interested behavior in water policy in order to ensure economic growth and "political stability". Therefore, local actors, rather than national ones, will remain the most influential protagonists in border water politics. This point is demonstrated in the next two sections which analyze water management in the two case studies cited above.

IV. Silao, Guanajuato: A regional Mexican Case of a Thirsty City Growing Drop by Drop

Silao, located in central Mexico, is a city that has been significantly affected by "regional integration", defined in this case as the industrial expansion of the city's economy within a large geographic region in central Mexico known as the *Bajío*. While Silao is administratively considered a municipality in the state of Guanajuato, it is also a central point of important overlapping geographic regions. Economically, it is part of an enclave inside the *Bajío*, whose great productive boom in the mining industry and agriculture over the last two centuries (known as the *epoca novohispanica*) significantly contributed to Mexico's economic development. The city also acts as the industrial locomotive of *El Bajío guanajuatense*, which first developed in the 19th century and later accelerated during the 20th century. Finally, Silao belongs to the hydrologic region of Lerma-Chapala-Santiago, specifically to the *Medio Lerma* subregion of the Cuenca *Lerma-Chapala* (Chapala Basin).

Economic Development and Environmental Risk

As part of the industrial centre of the *Bajío*, the city and municipality have been objects of decisions taken by governors, political leaders and businessmen for industrial development. Industries, such as Maseca (derived from corn) and infrastructure, like the airport, were established in the city and municipality by the state government, without consideration for the authority of local leaders, and without the consultation of local citizens. The local government of Silao has played the role of receiver of these decisions and it has had to confront new social demands (such as water shortages) which are the by-products of new industry in this territory, and the consequent process of urbanization. While industrialization has generally transformed the bases of the city's economy, the installation of a relocated General Motors (GM) plant, specifically transformed the city's economic landscape and created numerous social challenges. The 1.2 million sq. ft GM plant has almost 3,000 workers and it currently produces around 1,000 vehicles (basically Chevy suburbans and Cadillacs) per day.

One of these challenges is, in fact, a significant shortage of water resources. Industrial development has led to the overexploitation of the Lerma Chapala Basin. The Lerma-Chapala region has access to only 30% of the water resources in Mexico, while it represents approximately 70% of the national industrial activity. Moreover, it supplies water to around 15 million people (Caraza Tirado *et al.*, 1993). In the year 2000, the average flow of the Lerma-Chapala Basin was only 4,740 Mm^3 (Millon cubic meters), compared to an average recharge in the aquifers of 3,840 Mm^3. The result of this extraction was a deficit in the recharge of 900 Mm^3 (SEMARNAP, 2000).

The city of Silao is found in the middle of this basin and it has been impacted by a series of decisions that favor the supply of water to other cities, such as Mexico and Guadalajara (which are located at the extremes of the Basin) and to industrial complexes, like the GM plant, that exert significant influence inside the socio-political system, basing their arguments on their "economic importance". Thus, smaller municipal areas, such as Silao, are negatively affected by a water distribution system that is based on economic power.

Since the middle of the 20th century, Silao has been impacted by a series of productive transformations that played a central role in the water panorama of *El Bajío*. From 1940 to 1950 the state and federal governments generated two large works that placed the whole state of Guanajuato in a promising productive position: 1) the extension of the highway originating from Mexico City that created the Querétaro-Guanajuato-Aguascalientes industrial corridor; and 2) the construction of the large "Solís" and "Allende" dams on the Lerma River. These actions provoked agro-industrial activity for export and in the 1960s, several international producers of fruits, vegetables and different milk products established plants in the area. Among those that focused on international export, one finds transnational corporations, such as Ann

O'Brien and Bird's Eye (COPLADEG, 1991). The burgeoning industry was characterized by its specialization in the area of food production as the result of a rural modernization program called "the Green Revolution" during the 1960s-1970s. This program consisted in juridical and economic packages that supported the agricultural sector in this region. The impacts included economic growth as well as environmental degradation which lowered the quality of life in the area (Gonzalez, 1990).

There were, in effect, important consequences from these programs. The agro-industrial activity thrived on an excessive drilling of wells for the extraction of groundwater, which occurred parallel to a series of official decrees beginning in 1948, prohibiting the drilling of wells in the state (Cruz, 2000). The appearance of these emergency decrees demonstrates that water resources in the region were not being rationed adequately, which lead to a disparity between the economic growth of the 20th century and the optimization of the water use. While federal officials responsible for the management of water were indicating the presence of a latent risk of overexploitation, other federal institutions, with the support of the local governments in the area, were promoting the development of industry and export agriculture by increasing irrigation with groundwater throughout the region. The increased drilling of wells in Guanajuato was obviously incongruent with the prohibition decrees passed during this period. The process in its entirety illustrates how the decisions to promote uncontrolled economic and industrial development generate ecological risk.

Silao, in fact, is a growing city that needs water, but the central challenge in terms of governance is deciding how much. Also, leaders must decide how and when actors can acquire the water resources that they claim are necessary. Thus far, state and local governments have failed to respond to these questions in the face of continuous promotion of industrial installation and investment in huge infrastructure projects, such as the highways and dams cited above. Despite the prohibition statements from federal water officials, property and business owners with rights to wells have dictated drilling. For example, actors facing the extinction of their present wells often begin drilling negotiations before they receive reinstatement permission from the CONAGUA (the National Commission of Water) for their well, authorizing them to perforate land inside the same aquifer zone. Due to this behavior, the rural panorama of the city by the year 2000 was characterized by a continuous landscape of drilling machines for "the reinstatement of wells". In 2000 there were already more than 16,000 wells in the whole Guanajuato State. Probably only 25% of this total was registered by CONAGUA. By 2001 the number increased to 17,302 wells, despite the aforementioned "*vedas*" or prohibition statements (Maganda, 2004).

Each time actors drilled to greater depths, extracting water from the deep aquifer, at times without following the official technical requirements for drilling.

Because local officials have short time horizons, they often cede to these public pressures for increased water supplies. This led to a declared water emergency and an economic crisis within the city's water agency in 2003, caused by a lack of payments for water use. These political pressures come from many sources, such as multi-national corporations (notably General Motors), agro-industrial producers, and urban users. Influence is often peddled indirectly through specific institutions in the state government, such as the Office of the Secretary of Farm Development (SDA), the *Sistema Operador de Agua Potable y Alcantarillado de Silao* (SAPAS), and FES (municipal and state organization for the support of industry). Public debates concerning water use are barely gaining legitimacy as they often contradict accepted practices such as excessive drilling, overexploitation of aquifers, and the over-distribution of water resources. Political pressures and unplanned growth have led to impoverished discussions concerning the local water supply. This problem is further accentuated by a lack of institutional unity amongst water officials in different management agencies.

Water Management across Borders in Central Mexico

The process of regionalization in Silao has affected the city in many ways. Not only has the metropolitan area been transformed from a small agricultural community into an internationally recognized industrial conurbation, but the governmental bases of the city have also been significantly affected by the development of regional institutions. This has been the case in the arena of water management as decentralization and state challenges to national decision-making authority have fundamentally altered water distribution in the *Bajío*. Silao belongs to the Lerma-Chapala hydrologic region, where different institutional actors compete for supremacy in water allocation matters, often to the detriment of local communities, such as Silao.

The Lerma-Chapala Basin is one of the biggest in Mexico as it serves five states. The basin is characterized by: a 54,300 km^2 geographic range, a direct user population of 11 million people plus an indirect user population of 5 million, economic activity that represents 9% of Mexico's GNP, 795,000 ha of irrigation land and drinking water pumped out of the Chapala lake for 2 million people in Guadalajara (Mollard, 2005). The basin is governed by a Basin Council which is composed of the Governors of the states that draw water from it and it is managed by the CNA (*Comisión Nacional del Agua*). This system has largely failed in creating long-term agreements because of inter-party conflict and the pressure placed on leaders by economic lobbies and urban users to negotiate the largest amounts of water for their own states. The current President of Mexico, and former Governor of Guanajuato, Vicente Fox is famous for his phrase: "Starting today, not a drop of water will leave Guanajuato" which clearly demonstrates the interstate competition for the basin's resources. Moreover, political divisions between Governors from the PRI (*Partido Revolucionario Institucional*) and PAN (*Partido Acción Nacional*) have further weakened the efficiency of this institution.

Since 1989, the CNA (National Water Agency) has been the highest authority in water management throughout Mexico. The agency has been responsible for the creation and presentation of national water programs, and the planning of water resources in their regional contexts, i.e. within their basins. In 1993 the CNA reported a hydraulic imbalance in the Lerma-Chapala Basin's underground water, where the average annual flow of 1,364 Mm^3 decreased due to a demand of 1,557 Mm^3, which signifies a deficit of -193 Mm^3 per year. The lack of water to satisfy the Region's needs provoked difficulties for consumers due to contamination levels that limited its productive use.

Sine that time, some states found in the Lerma Chapala Basin have attempted to wrestle control over water management from the federal government directly challenging the CNA's sovereignty in this area. The first state to do so was, in fact, Guanajuato, where Silao is located, thus, making this state a critical case. In 1991, the State Water and Sanitation Commission of Guanajuato (CEASG) was created as a decentralized public entity for the provision of potable water, sewage removal, and sanitation. Until late 1995, the main function of the CEASG was to coordinate and execute potable water and sewage programs in urban and rural areas. In its first four years, it had two directors. In May 1995, Vicente Fox named Vicente Guerrero, then President of the Council for the System of Water and Sewage of Leon (SAPAL), as the new director of the CEASG. Guerrero implemented institutional changes which generated a broader focus for the organization. This permitted CEASG to more effectively tackle state water problems.

Among the technical changes made by Guerrero, one can identify measurement as one of the most significant. Previously, all of the basin states relied on the CNA for measurement information regarding water levels in the basin. One of the CEASG's first acts was to commission independent measurement so as to guarantee autonomous sources of information. This act was not only a challenge to the CNA's authority, but also a criticism of its efficiency and accuracy. Since 2000, the state has passed its own water law and established a State System of Water

Planning and Management. In institutional terms, this system included: the collection of documentary information (foundation of a documentation center), an area for the completion of basic hydraulic and socio-economic studies, the integration of information systems and generation of simulation models, and the definition of broad lines of action in water policy. The state also introduced plans for the monitoring of the drilling of wells, the social management of water, and the promotion of a "culture of conservation" among the local population.

Currently, the State Water Commission of Guanajuato, the CEAG, is the leading authority in all types of water use in Guanajuato. It is the institution to which municipal leaders and water users in Silao refer for the attainment of complementary water resources. Despite this fact, the agency cannot be considered completely autonomous because it works with a regional institution – the Lerma Chapala Basin Council. When local actors ask for increased amounts of water resources, CEAG representatives discuss the request and take it to the Basin level, where decision-makers focus on state needs, with little understanding of the pressures felt by local leaders and public servants in small cities, such as Silao. The local political leaders (from City government and the SAPAS in this case) present their arguments to CEAG officials describing with pride the economic importance of the municipality of Silao, its infrastructure projects, industrial parks and public works. They emphasize the city's rapid and historical industrialization process. Currently, the city is developing a "globalized" vision of economic expansion that is potentially viable but not guaranteed without increased access to water resources. This places increased pressure on officials to maximize their allotment from the Lerma Chapala aquifer, thus leading to a potentially disastrous situation. Because of the institutional confusion over water management and the competition between the various agencies erected for the preservation of this resource, there are few institutional constraints on local leaders to prevent them from following dangerous development strategies. While the case of Silao is domestic, it reflects the current situation found along many international borders, such as the US-Mexico frontier. This is the focus of the following section.

V. The US-Mexican Border: San Diego Water Needs and the Transfer Agreement with Imperial Valley

The US-Mexican Border is a 3,200 kilometer boundary of contrasts and disparities which presents an excellent "temperature indicator" of bilateral relationships examined by numerous scholars of water politics. In terms of the environment, this border region is characterized by large deserts, arid land and limited rainfall, in combination with multiple water uses that cause the degradation of water resources (i.e. golf greens, *maquiladoras*, and human consumption). On both sides of the border, cities and rural communities are facing unprecedented stress on limited water supplies as the result of both competition for water and the presence of persistent drought for many years (Gleick, 2002). The border region faces unequal distribution of water and poor infrastructure. Mexico depends on the same water that is badly needed in California, Arizona, Nevada and Texas. In the Mexican border region, the greatest need is for water and wastewater infrastructure in urban areas where sewer systems have exceeded their useful life and require rehabilitation. Less than half of the wastewater is treated. In most cities half of the piped water is squandered because of leakage and more than half of the irrigation water is lost to evaporation or seepage (Rosenblum, 2002). Like the Lerma Chapala Basin, water management in this area does not meet the needs of local residents and it does not guarantee future supplies of this fundamental resource.

Water Needs along the US-Mexico Border

A major problem facing the Colorado River is that it is over-allocated. More water is legally apportioned ('paper water') than actually flows ('wet water') (Cohen, 2002; and Gleick, 2002). Over-allocation results from the faulty average hydrology capacities established in 1922. In addition, the upper basin states, primarily Utah and Colorado, have not fully developed their shares, enabling lower basin states, particularly California, to overuse their entitlement. In accordance with the Colorado River Compact, the upper and lower basin states are entitled to the exclusive beneficial consumptive yearly use of 9.25 Bm^3 each. An option is granted to the lower basin states for an additional 1.23 Bm^3 for beneficial consumptive use. The 1929 California Limitation Act restricts California's annual consumptive usage to 5.43 Bm^3, plus not more than one-half of any surplus water not apportioned by the compact. For its part, the Metropolitan Water District (MWD) explains that California's annual use of the Colorado River water has varied from 5.55 to 6.41 Bm^3 over the last decade. The current use of up to 6.41 Bm^3 per year stems from the existence of surplus conditions and the availability of water unused by Arizona and Nevada (MWD, "The River").

The pressure on California to limit water use has generated concerns. The development of domestic water supplies was considered to be the state's 'highest and best use' of water, followed by agricultural usage (Davis, 2004). California had grown quickly, importing water from distant locations to meet its needs. According to Michael Cohen (2002), water in the western US developed under institutions designed to encourage settlement and consumptive, off-stream use. The state's over-usage and high consumption have caused what Davis labels "clash of the titans' style water fights".

Particularly, the city of San Diego has witnessed important growth since World War II, becoming a center for military production and a premier naval port. Since then, San Diego County has become the state's third largest, after Los Angeles (LA) and Orange counties, and it is more populated than twenty-two states in the nation (San Diego Book of Facts), with nearly three million inhabitants and growing water needs. To face these challenges, San Diego water authorities (inspired by the effectiveness of Los Angeles' hydraulic engineers "water barons"), began importing water over one hundred years ago (Carle, 2000). Currently, anywhere between 75 and 90% of San Diego County's water is imported (San Diego Book County Water Authority website). Like LA and its water transfers through canals from the Colorado River, San Diego became famous for bringing water from distant sources. A principal concern facing the region today is this reliance on imported water.

While appropriate technology has been necessary for long-distance water transfers, political power sufficient to negotiate investments, build dams, and get required allocations of water has been the key to assuring resources, particularly in areas with shortages. These deficits and the over-allocation of imported water pose a threat to the vitality of Southern California, prompting the region to look elsewhere, or internally (i.e. to irrigation districts) for solutions. The main complication in this case is not the growing population and the increasing demand, but the fact that local officials have been boldly promoting urban development and maximizing water allocation. Despite the fact that San Diego County residents overwhelmingly support the diversification of the area's water supply (San Diego County Water Authority: http://www.sdcwa.org/news/07224publicopinionpoll2004.phtml), local officials insist on importing water resources.

In 1947 the San Vicente Dam in San Diego received the first water from the Colorado River and with that this city's dependence on the Colorado began. San Diego's water situation became highly dependent on Los Angeles, once water was imported from the L.A. aqueduct and the Metropolitan Water District, and it was bought from the Colorado River Aqueduct (see map 3). After 1947 San Diego's growth projections were often artificially inflated for water negotiation purposes. There is an inextricable connection between water and opportunity structures in negotiations. Demand for water has generally been based on what a community can negotiate rather than its needs.

Despite success in water negotiations, San Diego suffered a drought that lasted from 1987 to 1991 and created tense competition under water allocation agreements along the Colorado River and forced San Diego water-seekers to re-focus their attention on local sources. 1991 was an especially hard year for San Diego due to water reductions from the Colorado River, Mono Lake and Owens Valley, demographic, commercial and industrial expansion, as well as military demands. In fact, the SDCWA serves more than three million people populating more than 518 km² of developed land. In spite of the negative environmental predictions for San Diego's water future, local economic and political leaders actively negotiated expansion of local water sources. They have improved and increased the desalination plants and ground water extraction, as well as developed a water transfer plan for the near future (SDCWA web site).

The International Impact of Local Water Management

San Diego's effort to diversify the sources of its water supply, long ago lead by SDCWA, has forced leaders to re-evaluate agricultural-urban relationships in water distribution. For years, San Diego's water leaders had looked with interested eyes towards the Imperial Valley Irrigation District (IID). This idea has developed quickly since the first announcement of a "probability of a water transfer agreement in 1988", up to the actual signing of the agreement in 2003. The plan has passed over important socio-economic factors, such as the higher price of this water for irrigation, and the opposition of farmers and environmentalists, in order to include this water transfer as a component of the state's conservation strategy.

Allocation problems in this region have also crossed numerous borders. The Imperial Valley in California and the Mexicali Valley in Baja California (Mexico) have been centers of development in the lower Colorado River Delta due to irrigation. The 1944 treaty formally allocated Mexico 1.85 Bm³ from this delta (one-tenth of the river's estimated flow). But the treaty also stipulated that during years of drought any shortfall required to meet Mexican rights would be substituted by equal quantities from the basin states. Vagueness in the treaty, caused by failure to define "extraordinary drought" and "quantities distributed" has lead to conflict over claims to the river. Mexico cannot participate in water management debates and it currently receives insufficient amounts allocated by the outdated 1944 treaty.

Once the proposed transfer's environmental impact statement was finished and environmentalist opposition relaxed, Southern California water agencies signed the final accord in October 2003 with the following terms: SDCWA will receive up to 246 Mm³ per year of IID water, and will be responsible for lining the All-American and Coachella canals with the State of California obligated to pay US $235 million. In return, SDCWA will receive 95 Mm³ per year from the All-American canal for 110 years (Walker, 2004). Up to US $300 million will be made available for socioeconomic and environmental costs in Imperial Valley, including Salton Sea restoration.[2] The agreement, however, does not specify from where this money will come, and there is no official study of the socioeconomic impact of this water transfer on neighboring Mexican cities, which were not included in the third-party report. Besides this environmental struggle, the California-US conservation plan, taking shape just kilometers from its wheat fields, means less water will percolate through the sandy soil into Mexicali's underground water supply, threatening crops that are the lifeblood of Baja California's richest agricultural region.

Without any established legal binational document governing the AAC groundwater delivery, approximately 98.6 Mm³ annually were allocated, to Mexico under the 1944 treaty, on top of 1.85 Bm³ surface water, filling Mexicali Valley's underground aquifers. This interdependence between Imperial and Mexicali Valleys, has allowed agricultural, economic and urban development on the Mexican side for over 60 years. Mexicali is now the third-largest Mexican border city, with a population close to one million. The filtered water has the highest quality on the northeast side of the Mexicali Valley. It has traditionally been used to develop agriculture. The lining of the AAC would greatly limit groundwater recharge into Mexico.

This problem is a reality that works against the spirit of regional integration. Since the 1992 North American Free Trade Agreement (NAFTA), the Mexican-US border has dominated public agendas. Thus, management decisions are no longer sovereign (Mumme, 2005). While, their execution remains national, their effects are binational.

VI. Challenges to Institutional Water Frameworks and the Need for New Water Management Structures in North America

By presenting these cases of study together, I have attempted to show that water politics in shared basins are dictated by the same mechanism, whether the basins cross sub-national or national borders. Political actors, in both the Lerma Chapala basin and the Colorado basin, compete within a framework created by institutional and legal constraints. Water politics in both cases, are characterized by significant

---

[2] Under this agreement, there is a 200,000 AF flow reduction into the Salton Sea. Restoration plans must take into account the lower lake levels and the higher salinity levels that could result (Davis, 2004).

structural difficulties related to governance, including ecological problems and socio-political risk caused by overexploitation of resources that extend beyond the jurisdictions of management agencies. Furthermore, in both the Lerma-Chapala and Colorado basins, institutional bodies compete for increased amounts of water resources instead of collaborating on innovative ways to manage water and guarantee its availability for the future. This, of course, raises important questions related to democracy and governance in border areas because current management strategies contribute to a dysfunctional system. Those agencies charged with regulating self-interested behavior actually violate the spirit of their own mandates by following political incentives that are self-motivated. Thus, a fundamental question that needs to be addressed is: What can be done to improve water regulation in border regions? In both of the cases presented in this chapter administrative reform is definitively necessary for more efficient water regulation. However, water management could also be improved if the current structures governing the water basins (Lerma-Chapala Basin Council and the International Boundary Water Comission) were made more transparent, more democratic, and more effective. These reform objectives reflect the three most evident structural problems in North American border water management systems: 1) the lack of public information regarding water resources and decision-making for their distribution, 2) the lack of public oversight of water management agencies, and 3) the inability of basin councils or commissions to limit the self-interested behavior of local actors.

In the Mexican case presented above, the CEAG's efforts to decentralize some political mechanisms in water management are notable. Recent reforms, such as the creation of independent studies of aquifers, the monitoring of wells, the implementation of state mathematical models for the hydrodynamic application of water flows, the state distribution of economic resources for water infrastructure in each municipality, and public conservation campaigns are significant. All of these actions may seem logical, especially when compared to water policies implemented by US states, but they actually represented a major political challenge within the Mexican water regulation system and they created strong tensions with the federal government. Since 1995, and particularly after 1998, the CEAG has been promoting many changes within its own institutional structure, as well as in its role in the state executive in order to improve transparency and make water regulation more democratic. CEAG is the first Mexican state water agency to have its own water planning structure and Guanajuato is the first state to include water management in its executive cabinet along with traditional arenas, such as governance, economic development, health and education.

Of course, institutional reform has been smoother than socio-economic change. It has not been easy to reconstruct a water institution with plural objectives in the face of self-interested agro-industrial and industrial leaders that have been used to receiving water in desired terms and quantities. Moreover, economic leaders have opposed reform because they fear that it will slow the rhythm of development in the state and urban users have complained about the new governmental pressure to pay for water services more frequently. Thus, water reform touches numerous arenas of democratization in Mexico as it affects economic development, institutional power, and civil society.

For these reasons, specific objectives need to be followed to ensure effective border water governance in Mexico. First, water management at the state level needs to be de-politicized and a balance needs to be found between public service, social pressures, and economic develop-ment. CEAG has begun to address these issues by sharing accurate water information, developing long term strategies for water manage-ment, and promoting new water conservation programs. Second, a clear national water management structure needs to be developed in which the responsibilities and authority of state and federal agencies are clearly demarcated. This will attenuate and hopefully eliminate competi-tion between Mexican states and the CNA which should collaborate in water management rather than undermine the practical long-term distri-bution of this resource. Third, the Lerma Chapala Basin Council needs to be made more open to citizen participation and oversight. Due to a lack of transparency, state officials in the council follow the interests of their political parties rather than their citizens. One way to address this problem is to elect public members who are unaffiliated politically.

As demonstrated above, water management along the US-Mexico border suffers from similar structural deficiencies, even if the level of governance is different. Like the domestic Mexican context, institu-tional confusion regarding authority and responsibilities characterizes water regulation along the US-Mexico border. This problem is espe-cially disturbing because international protocols exist which could solve this problem if they were implemented. For example, the Helsinki Rules on the Uses of the Waters of International Rivers establish some interna-tional legal guidelines for an equitable utilization of the waters of an international drainage basin, including surface and underground waters flowing into a common terminus. The US government still does not recognize the regional concept of an "international water basin" and it only addresses surface water issues explicitly included in the 1944 treaty. Because of this US attitude, there is little that the binational

IBWC/CILA, can do to protect Mexican water rights in the Colorado Delta. Thus, like the Mexican context, institutional clarity needs to be developed. Specifically, the IBWC/CILA needs to be empowered with institutional means to enforce water rights on both sides of the border. Moreover, the inclusion of public members from both the US and Mexico would improve the transparency of this agency's work.

In order to create an equitable situation at the border, despite the inequity of economic power, international arbitration is necessary. The Mexican part of the International Boundary Water Commission (CILA) has the elements to present a formal complaint to the International Court of Justice (ICJ) as an "environmental frontier dispute against the US for not respecting the aforementioned international Helsinki rules on the uses of the water in the binational Colorado River basin". Of course, the ICJ will force both countries to assume their respective responsibilities, and its decisions also depend on the willingness of a country to recognize its jurisdiction. While it may be unlikely that the US will follow ICJ rulings, the involvement of an international body could create enough political pressure to influence US officials and push them to open negotiations with Mexico on equitable border water practices. Cross-border equity, in fact, is the foundation of efficient border water governance because it reduces social and political conflict and it ensures longer time horizons in water management decision-making. In order to operationalize this norm, it is necessary to create effective transboundary water agencies that are truly based on the notion of regional integration in practice, as well as in name.

References

Aboites, Luis. 1988. *La irrigación revolucionaria. Historia del Sistema Nacional de Riego del Río Conchos, Chihuahua (1927-1938)*. México, Secretaría de Educación Pública-CIESAS

———. 1994. "Irrigación, desarrollo agrícola y poblamiento en el norte de México (1925-1938)", in Viqueira Landa, Carmen y Torre Mora, Lydia eds. *Sistemas hidráulicos, modernización de la agricultura y migración*. México: Colegio Mexiquense, A.C./Universidad Iberoamericana, 1994, p. 431-459.

———. 1995. *Norte precario. Poblamiento y colonización en México (1760-1940)*. México: CIESAS, El Colegio de México, 1995.

———. 1998. *El agua de la nación: una historia política de México, 1888-1946*. México: D.F., CIESAS.

Aguilar Barajas, I. and Mumme, S. 2003. "Managing Border Water to the Year 2002", in Michel, S. ed. *The US-Mexican Border Environment – Binational Water Management Planning*. San Diego State University Press: San Diego, 51-93.

Anton, Danilo. 2003. "Saciando la sed planetaria: los problemas del agua en el fin del milenio", in Ávila García, Patricia ed. *Agua, Medio Ambiente y Desarrollo en el siglo XXI.* Zamora: El Colegio de Michoacán, SEMARNAT, IMTA, Secretaría de Urbanismo y Medio Ambiente de Michoacán.

Bennett, Vivienne. 1995. *The Politics of Water: Urban Protest, Gender, and Power in Monterrey, Mexico.* University of Pittsburgh Press.

Bennett, Vivienne and Herzog, Lawrence. 2000. "US-Mexico Borderland Water Conflicts and Institutional Change: A Commentary". *Natural Resources Journal*, Vol. 40, p. 973-988.

Beriain, Josexto (Comp.) 1996. *Las consecuencias perversas de la modernidad. Colección Autores Textos y Temas Ciencias Sociales.* Barcelona : Editorial Anthropos.

Berry, Kate A. 1997. "Of Blood and Water". *Journal of the Southwest.* Vol. 39, No. 1, Spring 1997.

Blatter, Joachim and Ingram, Helen eds. 2001. *Reflections on Water: New Approaches to Transboundary Conflicts and Cooperation (American and Comparative Environmental Policy).* The MIT Press, 1st ed. (22 January 2001) ISBN: 026202487X

Brown, C., Wright, R., Lowery, N., and Castro, J.L. 2003. "Comparative Analysis of Transborder Water Management Strategies: Case Studies on The United States – Mexico Border". *The US-Mexico Border Environment: Binational Water Management Planning*, San Diego: San Diego State University Press.

Brown, F. Lee, and Ingram, Helen. 1987. *Water and Poverty in the Southwest.* Tucson: University of Arizona Press.

Davis, Mike. 2002. *Dead cities, and other tales.* New York, W.W. Norton.

Dobkowski, Michael N. and Wallimann, Isidor (editors). 2002 *On the Edge of Scarcity. Environment, Resources, Population, Sustainability, and Conflict.* Syracuse, NY: Syracuse, University Press.

Fernandez, Linda and Carson, Richard T. eds. 2002. *Both sides of the border: transboundary environmental management issues facing Mexico and the United States.* Boston: Kluwer Academic Publishers.

Hamilton, Cynthia. 1993. "Environmental Consequences of Urban Growth and Blight", in Richard Hofrichter ed. *Toxic Struggles, the Theory and Practices of Environmental Justice.* NY: New Society Publishers.

Ingram, Helen. 1990. *Patterns Of Politics In Water Resource Development.* Albuquerque: University of New Mexico Press.

Ingram, Laney, Gillilan, 1995. *Divided Waters: Bridging the US-Mexico Border.* Arizona: The University of Arizona Press.

Levi, Margaret. 1988. *Of rule and revenue.* Berkeley: University of California Press.

Maganda Ramírez, Maria del Carmen. 2004. *Disponibilidad de Agua, Un Riesgo Construido. Vulnerabilidad hídrica y crecimiento urbano-industrial*

en Silao, Guanajuato. México. (Ph.D. Thesis), México: Centro de Investiga-
ciones y Estudios Superiores en Antropología Social (CIESAS).

Mendez, Elizabeth, 1993. "La distribución del agua en Tijuana como factor de
marginalidad urbana". *Serie COLEF I, Frontera y Medio Ambiente.* Vol V.
El COLEF y Universidad de Ciudad Juarez, p. 111-135.

Michel, Suzanne. 2003. *The US-Mexican border environment: binational water
management planning.* San Diego, CA: San Diego State University Press.

Mumme, Stephen. 1993. "Innovation and Reform in Transboundary Resource
Management: A Critical Look at the International Boundary Water Commission, United States and Mexico". *Natural Resources Journal,* Vol. 33, Winter 1993 No. 1 (93-120).

———. 2003. "Revising the 1944 Water Treaty: Reflections on the Rio Grande
Drought Crises and Other Matters". *Journal of the Southwest.* Vol. 45,
No. 45, Winter 2003.

Mumme, S. and Brown, C. 2002. "Decentralizing Water Policy on the US-
Mexico Border". *Protecting a Sacred Gift: Changes in Water Management
in Mexico.* San Diego: Center for US-Mexican Studies, University of California.

Mumme, S. and Donna Lybecker 2005. "El Canal Todo Americano: perspec-
tivas de la posibilidad de alcanzar un acuerdo bilateral", in Sánchez Munguía,
Vicente ed. *El revestimiento del canal todo americano. Competencia o
cooperación por el agua en la frontera México-Estados Unidos?* Tijuana: El
Colegio de la Frontera Norte and Plaza y Valdes Editores.

North, Douglass. 1990. *Institutions, Institutional Change and Economic
Performance.* Cambridge: Cambridge University Press.

Nye, Joseph S. 2004. *Soft Power.* New York: Public Affairs.

Pombo, Oscar A. 2002. *Pobreza agua y Condiciones Sanitarias en la Frontera
Norte.* Tijuana: Colegio de la Frontera Norte.

———. 1999. *Water, Sanitation and Poverty in the Mexican Borderlands:
Considerations of Water and Sanitation Strategies Used by the Poor in Ti-
juana.* Doctoral dissertation, Department of Environmental Analysis and
Design, School of Social Ecology, University of California Irvine.

Simioni, Daniela. 1999. "Institutional capacity-building approach to urban
environmental governance in medium-sized cities in Latin America and the
Caribbean", in Inoguchi, Takashi, Newman, Edward, and Paoletto, Glen eds.
Cities and the environment: new approaches for eco-societies. New York:
United Nations University Press.

Spalding, Mark, 1999. *Sustainable Development in San Diego – Tijuana.* La
Jolla: Center for US-Mexican Studies, University of California San Diego.

Van Shoik, Rick, Brown, Christopher, Lelea, Elena, Conner, Amy. 2004.
"Barriers and bridges: managing water in the US-Mexican border region".
Environment January-February 2004.

Varady, Robert G. with Milich, L. 1998. "Managing transboundary resources:
Lessons from river-basin accords". *Environment,* Vol. 40(8): 10-41, October
1998.

Walsh, Casey. 2005. "Región, raza y riego: el desarrollo del norte mexicano, 1910-1940". *Nueva Antropología*, Vol. XIX, (64) Jan.-Apr. 2005: p. 53-73.

Weber, Max. 1978. *Selections in Translation*, ed. by W. G. Runciman. New York: Cambridge University Press.

Wescoat, James, L Jr., Halvorson, Sarah, Headington, Lisa and Replogle, Jill. 2002. "Water, poverty, equity, and justice in Colorado: a pragmatic approach", in M. Mutz, Kathryn, C. Bryner, Gary and S. Kenney, Douglas eds. *Justice and natural resources: concepts, strategies, and applications.* Washington: Island Press.

Websites Consulted:

http://www.undp.org/water/index.html Water Governance program (part of the UNDP: United Nations Development Programme).

http://www.giwa.net/ Global International Waters Assessment (GIWA).

http://www.iwra.siu.edu/ International Water Resources Association (IWRA).

http://uttoncenter.unm.edu/ Utton Transboundary Resources Center.

http://www.pacinst.org/ Pacific Institute.

http://www.nacla.org/ North American Congress on Latin America (NACLA).

http://www.transboundarywaters.orst.edu/ Transboundary Freshwater Dispute Database, Oregon State University Department of Geosciences.

http://www.internationalwaterlaw.org/IntlDocs/Helsinki_Rules.htm   Helsinki Rules on the Uses of the Waters of International Rivers.

http://www.icj-cij.org/ International Court of Justice.

PART II

BORDERS, ACTORS AND MOBILIZATION

CHAPTER 4

Boundaries in a 'Borderless' Europe

European Integration and Cross-Frontier Cooperation in the Basque Country

Zoe BRAY

I. Introduction

The launch of the European single market, the dismantling of frontier controls under the Schengen Treaty and enhanced inter-state relations have all helped to create new conditions for relations between neighbours across State divides. European integration involves a process of political, social and economic change in a geographically distinct European space (Wivel, 1998). As the mobility of people and goods become easier, so horizons broaden and mental and physical frontiers shift in ways that can be beneficial to sub-state national entities. With the increase of cross-frontier flows, the dominance of the nation-state idea diminishes and regions emerge as more important territorial units and political agents. Increasingly assertive in processes of economic and social development, regions are fast becoming contestable power centres, prompting competition for influence among local actors and between regions seeking advantage at a national and international level. In such a context, the opening up of frontiers can help state-less nations in border areas to forge closer links with 'compatriots' on the other side (Batten, 1995; Markussen, 2004), thanks to new opportunities for cultural and political affirmation that may pave the way for renegotiation of established boundaries. As a source of values of democracy, social and institutional organization and economic development (Sasse, 2004; Radaelli, 2000), Europe provides a forum for the discussion of new ideas and approaches to space, political authority and territorial development (Badie, 1995; Paasi, 1996).

Drawing on these new opportunities for debate, state-less nationalists in power in some regions have begun to explore new approaches to sovereignty and the search for enhanced self-determination (Keating, 2004). The scope for change remains nonetheless dependent on institutional arrangements at the broader state level. Despite EU emphasis on decentralization and regionalization, member states continue to hold the ultimate word on their 'interior' affairs (Keating, 1998). As a result, the possibilities for regions and state-less nations to affirm themselves in the new context of global and European integration depend on their particular institutional set-up within a given State territory and on their relationship with central government. While the borders of some regions are defined on a basis of clear cultural boundaries, those of others reflect economic or political considerations. Where economic, political and cultural identifications coincide, the cohesiveness of the region's profile is reinforced. The most powerful regions have tax-raising abilities that enable them to fund their own budgets independently of central government while at the same time enjoying a high degree of control over areas of public competence. Variations in the status and contextual situation of regions across the EU are reflected in the varying degrees to which inter-state frontiers can be said to have disappeared and regions to have affirmed themselves as political units.

This article confirms the general claims of this collection of articles, that cross-border policy-making does not always create expected outcomes. Border politics is a complex process of relations between center and periphery in which supranational initiatives affect the bargaining positions of border communities. Even though economic globalization and political regional integration have exposed communities to international forces, local actors have an influential role to play, while all the same having to wrestle with external political influences. In this article, we shall examine here local and regional mobilization in the Basque Country in the context of European integration. Located in the territories of two different States, France and Spain, the Basque Country is divided into distinct territorial units with different regional set-ups. Various terms are used to describe the region, Basque Country in English, Euskal Herria and Euskadi in Basque, País Vasco in Spanish, and Pays basque in French, each with different political and cultural connotations. According to Basque nationalists, the Basque Country covers the territory occupied by seven historic territories, or provinces: Araba, Bizkaia, Gipuzkoa and Nafarroa, or Navarre, on the Spanish side of the frontier, and Lapurdi, Behe Nafarroa and Xiberoa on the French side. Euskal Herria and its variants refer to the Basque Country as a linguistic and cultural territory, ignoring French and Spanish institutional frontiers. Not all of the area supposedly forming the Basque territory today can clearly be identified as Basque, however. Much of Navarre is Spanish-, rather than Basque-speaking, and its southern parts have little that is Basque about them. In the provinces of Bizkaia and Araba only a small proportion of the population is Basque-speaking. North of the frontier, the main city Bayonne has Gascon, as well as Basque, cultural roots.

Euskadi in its present-day form is an adaptation of the invented name Euzkadi, dreamed up by Sabino Arana, the founder of the first Basque nationalist party Eusko Alderdi Jeltzalea (EAJ),[1] in the late 19th century to refer to a notional Basque homeland. Today, it is used as the Basque name for the region known in Spanish as the Comunidad Autónoma Vasca, or Autonomous Basque Community, grouping three of the seven historic provinces of the Basque Country, Araba, Bizkaia and Gipuzkoa. In Spanish, the term País Vasco is commonly used to refer to this restricted territorial area. Navarre, meanwhile, reflecting its historically different relation with the Spanish crown, maintains its own territorial identity as an autonomous region with a similarly extensive array of administrative powers, despite the practice of Basque nationalists to include it in their references to Euskal Herria.

In France, meanwhile, the term Pays basque is used to refer to the part of the Basque Country in French territory. Only since 1997 have the 157 communes of the provinces of Xiberoa, Lapurdi and Baxa Nafarroa been recognised by French law as constituting a politico-administrative unit known as a pays, under the name of the Pays basque. Beyond that, the French Basque Country has no clear institutional status of its own. It forms part of a larger and culturally diverse territorial unit called the département des Pyrénées Atlantique, endowed with a limited set of competences, which in turn is part of an even larger and yet more culturally arbitrary region, Aquitaine.

At the central State level, France and Spain have very different interpretations of 'territory' and of the purpose and function of the region. Indeed, if we look across Europe as a whole, these two countries may be seen as polar opposites in terms of the degree of independence accorded to the region in relation to the central State. The different institutional characteristics of Euskadi, Navarre and the Pays basque, as parts of Spain and France respectively, have resulted in quite different identity constellations and, consequently, different political reactions to Basque nationalist mobilization.

In Euskadi, the regional government is presided over by an alliance of moderate Basque nationalists and Izquierda Unida, a post-communist party, under the leadership of Juan José Ibarretxe, a member of the traditional Basque nationalist party EAJ. One of his party's leitmotifs is a campaign, formalised in the so-called Ibarretxe Plan, to gain more power in running the affairs of the region, but also to expand Basque

---

[1]   Meaning Basque nationalist party.

nationalist influence beyond traditional regional and state boundaries so as to the broader territory of the Basque nation. In Navarre, a different historical tradition involving closer allegiance to the Spanish crown has limited the influence of Basque nationalism, which is concentrated in the northern Basque-speaking part of the region close to the frontier with France. The Navarre regional government is run by the Union del Pueblo Navarro, a right-wing formation close to the Partido Popular which was in power at state level until ousted by the Spanish Socialist Party in 2004. Basque nationalists, of different ideological traditions, have only eight out of fifty seats in the regional parliament.

In the Pays basque, electoral representation of Basque nationalists is even weaker. Though they have approximately 10% of the popular vote, they have only one representative in the Conseil Géneral of the départe-tement des Pyrénées Atlantiques and none in the regional assembly of Aquitaine. In October 2005, 64% of mayors in the Pays basque voted in favor of the creation of a Basque *département*, marking the latest development in a campaign for institutional recognition of the Pays basque dating back to the late 18th century. Supporters of this campaign argue that devolution to this level would enable actors in the Pays basque to react more effectively to the local challenges posed by European integration through specific cross-frontier initiatives. Local members of French Statist parties on both the left and the right wings have expressed some sympathy with demands for the creation of a *département du Pays basque*. But this has not been followed through at the départe-mental and regional levels. Instead, French authorities responded in 2000 by creating a so-called *Convention Spécifique Pays basque* to channel the demands of local business, social and political actors in cooperation with representatives of the state, regional and départemental authorities. Although following the lines of "new governance", involving close dialogue and collaboration between institutions and civil society, (Urteaga and Ahedo 2004), this initiative can be seen as fundamentally a way for French authorities to avoid granting the Pays basque its own institutional status.

The slowness of French State and regional actors to engage with European notions and take advantage of funds for local development and cross-frontier cooperation in the Basque Country[2] is partly due to reluctance to engage too actively with Basque nationalists combined with fears of economic domination by the more powerful Spanish regions. But it also reflects a continued lack of legal tools common to both France and Spain to support cooperation at a decentralised level.

---

[2] Bray, 2004 & 2006; Lurraldea, 2005; Laborantxa Gambara, 2005; Enbata, 2005; Lamassoure, 2005.

Aquitaine, Navarre and Euskadi talk grandly about joint ventures such as trans-regional infrastructure and twinning arrangements between towns. In October 2005, these and other regions on either side of the state frontier were invited for the first time to take part in a high-level Franco-Spanish meeting in Barcelona on the topic of international transport infrastructure. However, representatives of sub-regional levels of government, such as the département des Pyrenées Atlantiques, were not invited to participate, and the initiatives discussed by regional and State representatives are contested by local actors in the Pays basque on the grounds that they take no account of the needs and demands of the local population, and that they are detrimental to local sustainable development.

II. Euskadi, Navarre and the Pays basque in the Context of State Politics

The differences between attitudes on regional and local identity observable between France and Spain reflect the different constitutional arrangements and histories of the two countries. In the second article of the 1978 Spanish constitution, communities with historically anchored identities are recognized as 'nationalities' within the unity of the Spanish 'nation', a concept resulting from an amalgam of different regional attachments developed since the Middle Ages of which Castile is considered by many as the 'core'. Spain was one of the first European countries to establish a unified nation by bringing together a loose confederation of kingdoms and statelets under one king. Although Navarre and the Basque provinces kept a high degree of autonomy under a so-called foral system of legal and financial privileges codified in what in Spanish are known as the fueros, centralizing trends in the 19th and 20th centuries led to the creation of a unified Spanish State. Under the 36-year dictatorship of General Franco regionalism was discouraged, with regional development programmes mainly confined to large infrastructure projects directed from the centre. After Franco's death in 1975, however, the regional territories were able to reassert a sense of distinct identity.

In France, by contrast, regions are still far from having a similar degree of autonomous powers. The key centres of political power in France are still the central government in Paris, the *départements* and municipalities, notably large and medium-sized cities. This restrictive approach to sub-national government is a product of the 'Jacobin' ideology of the French State, opposed to regionalism because of its associations with the reactionaries to the 1789 Revolution and the subsequent formation of the Republic. The French Nation-State is based on the ideas of unity and centralism, with French national identity taking precedence over provincial attachments. In such a context, decentralization in France is far from implying regionalization, and even less so regionalism (Keating, Deschouwer and Loughlin 2003). In regional policy-making, French politicians talk of 'aménagement du territoire', dividing up French territory into small units for 'management' and 'development' purposes. With the establishment of the DATAR (Délégation à l'Aménagement du Territoire et à l'Action Régionale), the State ensures that it follows territorial development and regional politics very closely. Since 1997, DATAR includes a special branch dealing with cross-frontier relations, the so-called MOT (Mission Opérationelle Transfrontalière).

The devolution of powers to the regions has been matched by parallel devolution of powers to the smaller administrative units of départements and communes, and more recently to so-called communautés de communes, territorial agglomerations of various communes together. As the electoral constituencies of the regions, the départements also have a key role to play in regional planning and economic development. The result is tensions between the regions and the départements, especially where a département forms a culturally distinct unit.

Spain today is a State composed of seventeen autonomous communities or regions coinciding by and large with historical and cultural boundaries. Under Spain's 1978 constitution, all have in principle similar powers. Their distribution, however, is different for every community, as set out in their respective 'statutes of autonomy' (estatuto de autonomía). There is a de facto distinction between 'historic' communities (Euskadi, Catalonia, Galicia, Navarre and Andalusia) and the others: only the historic communities collect their own taxes, while the others receive allocations according to the 'transferred' government functions. Reflecting such differences, attitudes to nationhood and identification also vary. Castilians, for example, generally consider themselves to be Spanish first and Castilian second, while Basques and Catalans take an opposite view, identifying with their respective region first and with Spain only second, or sometimes even third after Europe. In Navarre, meanwhile, regional identification is strong, but so is identification with Spain.

Political relations between the regions and the central government tend to be conditioned by the dynamics of the parties in power. Bilateral negotiations over the division of competences between the State and the autonomous communities are easier when the same political party is in power both at the regional and central level or when the regional government expresses loyalty to the central State. The governments of those regions in Spain whose territorial frontiers coincide more or less with historical and cultural boundaries find it easier to mobilise a sense of regional identity. Euskadi and Navarre are good examples of territories in which political mobilization takes place at the regional level. Their regional governments provide a secure base from which the majority party or parties can exercise a clear range of powers and which can be contested by those in opposition.

Euskadi, a territory of 7,500 km^2 with a population of approximately 2 million, has long had a difficult relationship with the central State. Fiscal relations with the central government are regulated by a so-called Economic Agreement which takes its inspiration from the traditional foral system. While Euskadi passes on a share of the taxes that it collects to the central government under a system known as the cupo, its regional government has consistently demanded further fiscal autonomy, claiming that it needs to be able to compete in the European market and that the Spanish model of assistance to the poorer regions has been ineffective.

Such hawkishness is driven by the strong separatist element in Basque politics, lately represented by the left-wing party in the opposition EHAK,[3] and forced to the forefront by the violent activities of the armed separatist group ETA. The Basque government, formed out of a coalition led by the Basque nationalist party, EAJ,[4] has stopped short of advocating outright separatism. Responding to the aspirations of the nationalist component of its electorate, however, it has made a point of developing a notion of Basque identification among the region's inhabitants, by promoting Basque culture as modern and dynamic. A significant part of its annual budget of just over seven billion euros budget is devoted to Basque language education, interpretation and translation and bilingual signs. It operates its own police force, known in Basque as ertzaintzak, as well as television and radio networks in both Basque and Spanish, the region's two official languages, and it has responsibility for education, transport and urban infrastructure. The regional parliament of Euskadi, meanwhile, promotes a Basque identity card issued by Udalbiltza, a grouping of municipalities across the whole of the Basque Country. Such initiatives are designed to encourage a re-thinking of traditional boundaries by advocating a symbolic re-visualization of Basque territorial space as a unified whole despite the varying legal and political statuses of its components.

---

[3] EHAK (Euskal Herrialdeetako Alderdi Komunistak, meaning Communists of the Basque Country) is a surrogate of the banned left-wing separatist party Batasuna.

[4] Coalition members also include EA (Eusko Alkartasuna, meaning Basque Gathering) the moderate Basque nationalist party with a social democrat bent, formed after breaking away from EAJ in 1987, and the Post-Communist party EB (Eskerra Batua, meaning United Left).

In Navarre, by contrast, the regional government makes substantially less investment in Basque culture and language. The region, covering an area pf 10,931 km^2 with a population of approximately 570,000, collects most of its taxes, passing on a share to the central State under a system similar to that of Euskadi. Although the ruling UPN party is far from agreeing with the line of action of Spain's ruling Socialist Party, its adherence to the notion of Spanish Statehood means that it entertains a more collaborative relation with the central government. Support for Basque nationalism in Navarre is splintered among a range of political parties including, in addition to EAJ and EA, homegrown left-wing parties Aralar and Batzarre. The Basque language has official status only in those parts of the region which have a moderately significant proportion of Basque-speakers. Cooperation between the governments of Navarre and Euskadi is minimal, reflecting their differences on issues relating to Basque nationalism. In 2001, the two regions decided to end an arrangement whereby they shared participation in a common fund with the region of Aquitaine, in order to allow each region to manage its own relations with Aquitaine via separate funds.

France's system of regions was created in 1964 under the presidency of General de Gaulle, sometimes following the boundaries of traditional provinces, duchies and counties, and sometimes cutting right across them (Ardagh, 1982; Hoffmann-Martinot, 1999; Touraine, 1981). In the early 1980s, President François Mitterrand's Socialist government went a step further in decentralization with the creation of 22 regional authorities elected by universal suffrage. Regional councils made up of councillors elected at the level of the départements are presided over by a president with some of the executive powers once exclusively reserved to the préfet, the so-called civil servant who represents the State in each French territorial unit. Benefiting from the status of collectivité territoriale, previously reserved only to the départements and communes, they have responsibility for regional planning and economic development as well as the management of funds and subsidies. Other powers presently include infrastructure development, and more recently secondary and higher education, and professional training.

Only in a few French regions do cultural and administrative boundaries more or less coincide. Brittany broadly covers the same territory as the ancient Duchy of Brittany, for example, while Corsica, as an island, is a clearly defined regional entity. By contrast, the region of Aquitaine, despite taking its name from that of the ancient Duchy of Aquitaine, encompasses territories that were not part of the Duchy, including the Basque provinces of Lapurdi, Behe Nafarroa and Xiberoa. Covering an area of 41,000 km^2, more than five times the area of Euskadi and three times that of Navarre, Aquitaine is made up of five départements:

Dordogne, Gironde, Lot et Garonne, Les Landes and Pyrénées Atlantiques, With a population of 3 million, it has an annual budget supported by the State-region contract of approximately 970 million euros, less than one third of Navarre's annual budget and one seventh that of Euskadi's.

In addition to creating a zone of influence for its capital, Bordeaux, Aquitaine illustrates an approach to regional boundaries in France designed to avoid smaller areas from mobilizing around their respective historical cultural identifications (Keating, 1985 & 1998). These include both the Pays basque and a much wider area extending into the regions of Languedoc-Roussillon and Midi-Pyrénées, where a locally based regionalist movement promoting the langue d'oc claims sway over an area known by its militants as 'Occitania' (Hammel, 1998; Génieys and Négrier, 1998). Within Aquitaine, the Pays basque, with a population of approximately 264,000 in an area covering less than 3,000 km^2, is nested in the administrative unit of the département des Pyrénées Atlantiques, which it shares with the culturally and linguistically distinct province of Béarn.

Controlled by the centre right party UDF, the Conseil général of the département des Pyrénées Atlantiques has powers that complement those of the region but which sometimes lead to clashes. The département has its own chamber of industry and commerce and chamber of agriculture. Branches of the University of Bordeaux are based in Pau, the main town of Béarn and administrative centre for the département, and in Bayonne, the main town of the Pays basque, where a subsidiary départemental office was set up in 1994 in response to complaints of isolation on the part of local actors in the Pays basque. In recent years, the département has been active in the domain of Basque culture and language, principally through the Convention Spécifique Pays basque, launched in 2000 with the participation of the region of Aquitaine and the central State. The official webpage of the Convention talks of its two main objectives as the following: to 'ensure the integration of the Pays basque as a strategic part of the euro-regional space' and to 'maintain and consolidate its internal coherence'.[5]

The Convention confers no institutional powers, but it recognises the Pays basque as entitled to special financial support from the State for local cultural development. It marks a significant step in relations between local actors in the Pays basque and French government authorities in that it involves official acknowledgement on the part of the French government of the fact that the Pays basque has specific needs in terms of socio-economic, cultural and linguistic development. Nonethe-

---

[5]   http://www.lurraldea.net/.

less, in practice, participants still see a great need for diplomacy and tact in relations with governmental authorities. For example, references to the 'normalization' of the Basque language are still seen as threatening, with government authorities preferring to talk of 'aménagement' – or 'organization' of the Basque language. The Convention provides for discussion of the demands of local Basque activists to two councils, one made up of local associations and the other of local elected politicians. These, however, only have consultative powers, in a decision-making process ultimately dominated by the State and by regional and departmental authorities.

Despite differing perceptions of the Basque Country and its history, the political parties operating in the Pays basque all seize on elements of Basque identification to support their campaigns. Cultural elements once considered old-fashioned, primitive and backward-looking are now viewed as potential sources of political capital. In this vein, even the local branches of the mainstream national French parties have in recent years begun employing Basque terms once reserved for Basque nationalists. In their speeches and manifestos, they routinely mention Basque identity and the importance of promoting the Basque language and culture.

Such rhetoric is a response to the mobilization of local Basque nationalist militants. In the 1960s, neo-Marxist ideas of internal colonialism and associated accusations against the French authorities of keeping the Pays basque in a state of underdevelopment led to the birth of the armed group Iparretarrak as a 'northern' version of ETA. Today, this group appears to be dormant, although separatist sentiment is kept alive by the left-wing nationalist party Batasuna, which operates in the Pays basque despite having been made illegal in Spain, and the home-grown left-wing pacifist party Abertzaleen Batasuna. Local branches of EAJ and EA are also present in the Pays basque. In the late 1990s, these moderate nationalist groups joined forces with local actors, including associations and businesses concerned with the cultural and economic development of the Basque Country, under the banner of a coalition called Batera, to campaign for the dissolution of the current département des Pyrénées Atlantiques and the creation of a 'département du Pays basque'. In addition, Batera calls for the creation of other institutions that it considers should accompany a formally recognised and dynamic territory, such as a university, a chamber of agriculture, and official status for the Basque language.

Faced with the slow-moving and often indifferent approach of French authorities, several steps have been taken by French Basque militants in this direction over the course of 2005. Faced with the apparent indifference of the official chamber of agriculture of the département to the concerns of Basque farmers, a group of nationalist-minded farmers launched a local Basque chamber of agriculture, Euskal Herriko Laborantxa Ganbara, in January. Then, in October, 64% of mayors in the Pays basque voted in a self-organised referendum for the creation of a Basque département. Such actions demonstrate that the new mode of governance established with the Convention is far from satisfying local actors. So far, French authorities have responded in their customary authoritarian way, with the prefect taking the Laborantxa Ganbara to court and dismissing the vote in favor of a Basque département as null and void.

III. Cross-frontier cooperation

Cross-frontier cooperation is important for Basque nationalists because of the opportunities that it provides for the expression of the existence of a Basque nation. By enabling local actors to work together across the frontier in a manner eroding State barriers, it allows real powers to be exerted and barriers and boundaries to be overcome, realigned and reinterpreted. Through collaboration with Aquitaine and other European regions, Basque nationalists in power in Euskadi are pursuing their objective of asserting the existence of a modern European Basque nation, free of the control of the Spanish state and able to present itself on its own initiative in international relations and at the European level alongside other States. In a speech in November 2004 to celebrate an agreement with Aquitaine to strengthen the so-called Euroregion jointly created in the early 1990s, Mr. Ibarretxe spoke of the importance of cooperation for "healing the scars of history that made it hitherto impossible for us to work together".[6]

Such statements form part of a general context developed by the Basque nationalists in power in Euskadi which finds its clearest formal expression in the Plan Ibarretxe's proposals for a new political statute for the region. In essence, this program amounts to an attempt to redefine the relationship between Euskadi, Spain and Europe by re-examining notions of national and civic identity. But while the Ibarretxe Plan claims to be designed to contribute to efforts to end continuing political conflict in the Basque Country, accentuated by the existence of the armed group ETA, it has proved politically controversial. This is because it challenges the basis of Spanish State sovereignty as enshrined in the constitution of 1978 by demanding the right for inhabitants of Euskadi to vote on 'a new political statute' for the region. Even though

---

[6] *Le Journal du Pays Basque*, 14 December 2004.

it does not entail full-scale separatism, such a project could radically change the region's relationship with the Spanish State.

An important milestone in the process of closer cross-border cooperation was provided by the Treaty of Bayonne, signed in 1995 by representatives of the Spanish and French States to formalise their collaboration across the frontier. Thanks to this treaty, municipalities close to the frontier were enabled to create legal frameworks for joint initiatives, which they had not previously been able to do. Thus, the border towns of Hendaye, on the French side, and Irun and Hondarribia, on the Spanish side, came together under the Spanish legal status of a Consorcio, whereby public entities from different national legal contexts are able to join forces (Tambou, 1999). Numerous local projects have been launched under the auspices of this Consorcio, gradually winning popular support despite the initial reticence of many local citizens at what they saw as a breach of traditional boundaries (Bray, 2004).

Neighbouring villages on the border have since embarked on similar cooperation initiatives, but none have been so active as Hendaye, Irun and Hondarribia. Despite the openness of the frontier and increasing exposure to the values of communities on the other side, many people continue to feel an attachment to their own State national context and occasionally express mistrust of 'the other' on the other side of the frontier. This is evident in the slow progress of cooperation between the French Basque town of Bayonne and the Spanish Basque city of San Sebastian, launched with much pomp in the late 1990s under the banner of the 'Basque Eurocity', but which has since never really picked up. The official objective of this project is to work on common projects which would more effectively reflect the reality of the increasingly cross-frontier lifestyle of much of its population. But it has so far gone little further than the initial symbolic act. The lack of a common legal tool to enable this initiative to develop is no doubt part of the explanation for its slowness to advance, along with the institutional differences on either side of the frontier. However, at the root of the problem lies a lack of real political will, particularly on the French side, to engage in this project.

Against this background, the stated objective of the Ibarretxe Plan is to promote a 'Basque nation which englobes all of the Basque Country in contrast to the current territorial organization imposed by the Spanish and French States'. A Basque Community 'freely associated' with the Spanish State would be free to make further changes in its formal relationship with State government in the future, according to the principle of self-determination. Basque citizenship, based on Spanish citizenship rules, would be open to all residents of the Basque Country and to people of Basque ancestry outside, with a provision that nobody would be subject to discrimination on the basis of identification or non-identification with the (State-less) nation. While initially focusing on the right of self-determination for inhabitants of Euskadi, the Ibarretxe Plan foresees the possibility of inhabitants of Navarre and the Pays basque joining the process.

Following regional elections in April 2005 which showed a waning of support for the moderate Basque nationalists leading Euskadi's regional government, the Ibarretxe Plan has in effect been put on hold. It nonetheless provides a good example of an attempt by a government of a State-less nation to rethink traditional boundaries in the context of the new opportunities opening up in today's era of globalization. In this context, boundaries cover both territory – in the sense of delimitation of what is Euskadi and what can be considered the Basque Country – and administrative powers, as well as language and culture. With the Ibarretxe Plan, the mainstream Basque nationalist parties are trying to supersede existing boundaries by introducing new notions of where territories/areas of influence begin and end.

Meanwhile, the President of Navarre's regional government, UPN member Miguel Sanz, made his first official visit in July 2005 to the French Basque province of Behe Nafarroa in order to formalize budding relations with the president of the département des Pyrénées Atlantiques, Jean-Jacques Lasserre, a member of the centre right political party UDF (Union pour la démocratie française). Rather than promoting Basque nationalist ideals, however, this was done with Navarre's own particular regional needs in mind. In collaborating with like-minded leaders of the département des Pyrénées Atlantiques, without the involvement of local Basque nationalists, the leaders of the government of Navarre are seeking to strengthen their region's profile in the wider context of European integration, for example through the construction of a highway across the Pyrenees and through the Pays basque in order to gain access to the rest of Europe. The project has so far failed to win support from the Aquitaine regional authorities, but it is strongly supported by Mr. Lasserre, to the consternation of local actors in the Pays basque who argue that it would be counter to the interest of local inhabitants. While an increasingly large part of local civil society is against this project, Mr. Lasserre continues to promote it in his effort to build a special relationship between Navarre and the département des Pyrénées Atlantiques.

IV. Conclusion

This analysis enables us to identify three major obstacles to cross-frontier cooperation in the Basque Country: the absence of legal tools; the different administrative powers on either side of the Franco Spanish frontier; and the different nationalist perspectives in the three areas making up the Basque Country: the Spanish Statist politics in Navarre and in the Pays basque, in contrast to the Basque nationalist politics in Euskadi.

For the moment, the Consorcio grouping Hendaye, Irun and Hondarribia remains a special case where cross-frontier cooperation has been successful due to the strong support expressed by the mayors of the three towns. While each of the three mayors comes from a different political tradition – the mayors of Hendaye and Irun are Socialist, and the mayor of Hondarribia is a member of EAJ – their regular exposure to the lifestyle of their neighbours has made them sensitive to the need to work together despite their differences.

Elsewhere, however, cross-frontier relations have been limited, due to the different perceptions and ambitions of politicians in power in the Pays basque and Navarre and Euskadi. While the concept of the Consorcio works for Hendaye, Irun and Hondarribia, other kinds of cross-border cooperation, for example involving hospitals or businesses or schools and universities, are still hindered by a dearth of effective legal tools. Neither the State authorities nor the European Union have done much to remedy the situation. Indeed, on the French side of the border, State authorities have actually taken action in the opposite direction. A current example of is the decision by the préfet of the département des Pyrénées Atlantiques to force the municipality of Hendaye to enter into partnership with other neighbouring French Basque towns in the new territorial unit of the communauté de communes, a measure which threatens to hamper Hendaye's role in the Consorcio and so reduce effective cross-frontier cooperation. At the time of writing, the municipality of Hendaye is trying to resist the préfet's order in court.

Such instances are typical of the problems caused by the differing statuses of political and institutional actors on either side of the frontier.[7] On the French side, major decisions are mostly taken at the départemental and regional level, with actors representing the Pays basque

---

[7] In his wish to respect France's particular institutional composition, José-María Muñoa Ganuza, member of EAJ and representative of the Basque presidency for 'external affairs', recently declared his aspiration for the set-up of a structure in which the French State, the region, the département, the various collectivités locales and local associations would be brought together, to serve as an effective partner for collaboration with the Basque government.

enjoying little opportunity to express their views. In Navarre and Euskadi, by contrast, decisions at a regional level are taken by authorities representing culturally homogeneous territories and endowed with extensive powers. As a result, different types of cross-frontier cooperation are taking place at different levels. Between the regions, joint projects for infrastructure and transport planning are labelled as 'Basque' or 'European' at opportune moments. The governments of Aquitaine and Navarre use such labels when they appear beneficial in terms of image. This is the case for example of the ambitious Euro region project between Aquitaine and Euskadi, in which a ground-breaking agreement has recently been signed for the development of an ambitious logistics platform for transport. The government of Euskadi meanwhile tailors its use of such labels to its Basque nationalist aspirations.

Between the département des Pyrénées Atlantiques and the regions of Navarre and Aquitaine, projects concern infrastructure at a more local level. And finally, between villages, towns and communes, associations and political groupings on either side of the frontier, projects take on a more concrete nature, directly involving citizens. Numerous associations and business ventures are currently joining forces across the frontier at the very local level. Some of these associations benefit from the support of the Basque government and its devolved structures for extra emphasis on their Basque national potential.

In addition to entertaining institutional, economic and infrastructure relations with the region of Aquitaine and the département des Pyrénées Atlantiques, the government of Euskadi supports grassroots initiatives in the Pays basque as part of its Basque nationalist vision. Associations working in Euskadi for Basque culture, language, education and economic development have links with similar associations in the Pays basque, and many benefit from the financial support of the Basque government and various devolved or semi-independent structures in Euskadi such as Udalbiltza. But such actions are frequently controversial. In 2005, the Basque government, for example, expressed support for the Laborantza Ganbara. A public firm created by the Basque government is currently providing administrative services in assistance to the Laborantza Ganbara, while the building in which are located the offices of Laborantza Ganbara are provided for free by a Basque nationalist trade union from Euskadi. These contributions have been condemned by the local préfet as 'foreign' meddling.

The consequences are that cross-frontier cooperation in the Basque Country is far from producing a harmonious and homogenous space. On the contrary, actors wishing to engage in more cooperation often do not know who to approach and how, and remain frustrated by legal and

administrative complications. The main drivers of cross-frontier cooperation and territorial change in the Basque Country, at least as far as
citizens living in the border zone are concerned, result from the basic
conditions brought about by European integration, that is, the dismantling of border controls and free circulation across the frontier. Thanks
to the growing cross-border exchanges, inhabitants of frontier zones are
increasingly exposed to values and symbols traditionally existing on
either side of the frontier. While such cross-frontier living and exposure
to other values and symbols may not necessarily result in positive
engagement (Bray 2004), with time this may change as familiarity sets
in. This can already be noted in the increasing identification with
Basque values among inhabitants of Iparralde, for example.

This analysis reveals the new political struggles over territory in today's context of European integration. We observe that, while European
integration may represent an opportunity for the renegotiation of established boundaries, at the grassroots level of the Basque Country different factions struggle to refashion established boundaries and sometimes
also further acquired positions, instrumentalizing the European discourse in the process. European integration provides local actors with
possibilities for new modes of action and discourses. But these remain
articulated in local political struggles and are limitedly translated at the
institutional level. Border integration is the result of political bargaining
within a dynamic political landscape characterized by shifting opportunity structures.

References

Badie, R. 1995. *La Fin des Territoires. Essai sur le désordre international et
sur l'utilité social du respect*. Paris: Fayard.

Bray, Zoe. 2004. *Living Boundaries*. Brussels: PIE Peter Lang.

Bray, Zoe. 2006. 'Europe as Living Boundaries: the case of the Basque Country', in Strath, Bo and Persson, Hans-Åke eds. *Reflections on Europe. Defining a Political Order in Time and Space*. Brussels: PIE Peter Lang.

Chaussier, D. 1997. *Quel territoire pour le Pays Basque? Les cartes de
l'identité*. Paris: L'Harmattan.

Enbata. 2005. "Relations internationales de proximité". *Enbata*, No. 1895,
22 September.

Hoffmann-Martinot, V. 1999. "Les grandes villes françaises: une démocratie en
souffrance", in Gabriel, O. and Hoffman-Martinot, V. eds. *Démocraties urbaines: l'état de la démocratie dans les grandes villes de 12 démocraties
urbaines*. Paris: L'Harmattan.

Keating, Michael. 1998. *The New Regionalism in Western Europe. Territorial
Restructuring and Political Change*. London: Edward Elgar.

Keating, Michael. 2004. *Plurinational democracy: stateless nations in post-sovereignty era.* Oxford: Oxford University Press.

Keating, Michael and Bray, Zoe. 2006. "Renegotiating Sovereignty: Basque nationalism, European Integration and the Ibarretxe Plan". *Journal of Ethnopolitics*, Vol. 5, No. 4, November.

Laborantxa Gambara. 2005. *Rapport Natura 2000.*

Lamassoure, Alain. 2005. "Une ère nouvelle pour les relations transfrontalières". *Le Journal du Pays Basque*, Wednesday 26 October, p. 2.

Letamendia, F. 1998. *La Construccion del espacio vasco-aquitano. Un estudio multidisciplinario.* Leioa: Basque University.

Lurraldea. 2005. *Pays Basque 2020: Acte II du Project de Territoire.* Assemblée plénière 12 septembre.

Paasi, A. 1996. "Inclusion, exclusion and territorial identities – the meanings of boundaries in the globalizing geopolitical landscape". *Nordisk Samhallsgeografisk Tidskrift.* 23:3-17.

Radaelli, Claudio. 2000. "Whither Europeanization? Concept stretching and substantive change". *European Integration Online Papers.* 17 Oct.; 4(8).

Salamon, Lester M. ed. 2002. *The Tools of Government: A Guide to the New Governance.* New York: Oxford University Press.

Sasse, Gwendolyn, Hughes, James and Gordon, Claire. 2004 *Europeanization and Regionalization in EU's Enlargement to Central and Eastern Europe. The Myth of Conditionality.* London: Palgrave.

Tambou, Olivia. 1999. "Le Consorcio Bidasoa-Txingudi: premier organisme publique franco-espagnole de coopération transfrontalière". *Quaderns de Treball*, No. 32. Institut Universitari d'Estudis Europeus.

Touraine, A., Dubet, F. and Wievorka, M. eds. 1981. *Le Pays contre l'Etat.* Paris: Seuil.

Urteaga, Eguzki and Ahedo, Igor. 2004. *La Nouvelle Gouvernance en Pays Basque.* Paris: L'Harmattan.

Wivel, Anders. ed. 1998. *Explaining European Integration.* Copenhagen: Copenhagen Political Studies Press.

and across Austria's borders with the former Habsburg provinces. Following up on Scott's analysis of cross-border cooperation, this chapter examines how the economic integration efforts of governmental elites are actually embedded within a common border identity. It enquires how the changing border context revives national minority politics and whether this might facilitate cultural integration across borders or promote the development of the border region.

Situated along Austria's Eastern-most state border, Burgenland is the country's youngest region with a long history of changing national belongings and the highest diversity of minority cultures: the Burgenland Croats, the Hungarians, and the Roma. The several small, territorially dispersed communities have almost fallen victim to assimilation in the German majority population. Its changing history and geographic position along the border has also made Burgenland Austria's most socio-economically disadvantaged region. But the recent historic transformations in Europe, starting with the opening of the Iron Curtain in 1989, to Austria's EU accession in 1995, and EU enlargement in 2004 have recently brought new hope to the region. Burgenland has been acknowledged EU objective 1 status as well as large INTERREG programmes, turning it into Austria's largest receiver of EU regional funds. The regional government has started to promote the multinational border heritage as an important capital for regional integration into the European market. Particularly for the development of cross-border cooperation, the transnational experience and cross-cultural knowledge of Burgenland's national minority organizations might offer an important social capital for institution-building. On the other side, the governmental efforts to overcome political and economic gaps are limited by narrow territorial self-interest interest and the mistrust between the populations East and West of the former Iron Curtain.

Questioning the cultural basis of cross-border cooperation, this chapter shows how Burgenland's more autonomous regional development policy promotes and thus transforms its multinational border heritage. An introduction will embed the region of Burgenland in the context of Austria's national minority politics. The regional government's political marketing of the border heritage will be analyzed with regard to changes – real or symbolic – in the field of nationality policy. In order to understand whether the government's cultural strategy is based on a broader institutional consensus in the region, the next part analyzes the different minority representatives' interpretations of ethnicity in the changing territorial context. Questioning whether the region's cultural development strategy actually strengthens cross-border cooperation, the next step is to identify the winners and losers in the field of minority policy. Even in the absence of any relevant supra- or transnational options for minority policy, European integration provides complex challenges to the cultural and territorial boundaries structuring the domestic institutions.

II. Burgenland: a Multinational Border Region in Austria

After the defeats of the two World Wars, the remaining territories of the Austro-Hungarian Empire were reconstituted as the Second Republic of Austria, a federal state based on a Germanized idea of the Austrian nation. The Austrian State Treaty of 1955 anchored the protection of the Croatian and Slovene minorities in the constitution, thus proposing that the Allied founding forces had intended a multinational basis of Austrian citizenship. But over time, the multicultural heritage of the former imperial capital city Vienna and of the border regions Carinthia, Styria, and Burgenland became mostly excluded from the majority German-speaking public. At the same time, Austria's external protection status for the German speaking population in the Italian province of South Tyrol motivated the Austrian government's intense international engagement for the promotion of minority protection in Europe.

The preparations for EU accession in 1995 and the EU sanctions in 2000 drew Austria's domestic human and minority rights situation into the critical focus of European politics, The country's long borders with the new Eastern European member states made it one of the states most territorially exposed to EU enlargement. Differentiated from the old autochtonous minorities, the new migrants from kin-states in the former area of the empire pose a challenge to the Germanized nation that gives rise to xenophobic aggression and extreme right propaganda. But for economic reasons there is strong official commitment to the promotion of cross-border cooperation and Euroregions. In sum, Austria was one of the first cases to bring the nationality problems which Central Eastern Europe inherited from the Habsburg Empire into the EU, thus challenging the civic idea of citizenship from the inside of Europe.

Based on the consociational democracy of Austria's second republic, this conflictive historic heritage had turned into a comparatively harmonious majority-minority relationship and locked into a highly institutionalized minority protection regime. Austria's political spectrum, including that of minority politics, has been defined less by language groups, and more by party politicization and their respective assimilation strategies. While Austrian minority politics focused on the conflict between the Slovene minority and the German majority in the region of Carinthia, there are six acknowledged minority groups living in Austria, mostly close to the Eastern border. South of the historically multicultural capital Vienna, the Eastern-most region Burgenland is characterized by the highest diversity of territorially dispersed minority cultures. The Burgenland Croats were Austria's largest minority estimated between 30,000 and 40,000 people, the Hungarians about 25,000, and the Roma between 30,000 and 40,000 in 1991.[3]

Between the 10th and the 12th centuries Hungarian border guards were settled in today's area of Burgenland, developing diverse local traditions, later a stronghold of Protestant Hungarian nationalism against the Habsburgs. At the time Burgenland joined Austria in 1921, the Hungarian minority was characterized by deep socio-cultural divisions between the 'Magyaron' bourgeois and former bureaucratic elites, who only used Hungarian privately; the agricultural successors of the original low aristocracy, who were additionally divided into three religious groups; and the 'Beres', for whom the use of the Hungarian language represented their social inferiority as farm workers who recently immigrated. Enforced by three refugee waves from communist Hungary in 1945, 1948, 1956, these inner social divisions resulted in decreasing identification with the minority group and language assimilation.

Settling since the 16th century in parts of Western Hungary, South Western Slovakia, and South Eastern Lower Austria, the Burgenland Croats developed a separate high language in the 19th century. When Burgenland became part of Austria in 1921, the nationalization of the church-run schools polarized the Croatian minority between the Social-Democratic camp aiming at socio-economic integration into the Austrian market and the Christian-Democratic camp aiming at protection of the Croatian ethnicity. During the Second Republic established since 1955, this political integration into the national party spectrum turned Austria's numerically largest minority into an internally polarized and therefore 'silent minority'.

Moreover, five larger Roma groups live in Austria's territory today, namely – in order of their presence in the Central European German-speaking area: Sinti, Burgenland Roma, Lovara, Kalderash and Arlje. Coming from Hungary, the Burgenland Roma were the first to settle in Burgenland and in the towns of the Eastern part of Austria from the 15th century onwards. The Holocaust, survived by only 600-700 out of 7,000 Burgenland Roma, represents a dramatic memory for the older

---

[3]   As the survey of the more recent census of 2001 is considered politically questionable by many minority groups, the statistics of 1991 are still the official numbers used by most actors involved in minority politics. Taking into account the problems of quantifying minority belonging, these numbers combine surveys of everyday language use in the census of 1991 and self-estimates by the minority organizations to constitute a rough comparison (Statistik Austria: www.statistik.gv.at; www.initiative. minderheiten.at/; www.gruene.at/10bl/; February 2004).

population, parts of whom were extorted in the immediate post-war period as stateless people and offered compensation only since the 1960s.

Starting in the inter-war period and increasingly since the 1960s and 1970s, labor migration has forced an increasing number of people belonging to either of the three minorities to leave their bilingual rural communities and move to urban centers outside Burgenland, particularly Vienna. Recent immigration waves from the respective kin states have brought a cultural revival to the shrinking autochtonous communities in and outside Burgenland. Beginning in the late 1970s and 1980s but increasingly since the 1990s, the Burgenland Croats as well as the Hungarians and the Roma have achieved legal improvements, making Burgenland one of the most minority-friendly regions in Austria.

Due to its federalist constitution as well as to the historic heritage of the multi-national Habsburg Empire, Austria's present minority regime is characterized by a high degree of diversity among the administrative regions as well as among the local, regional, and federal levels. Historically, the nationalities question of the Austro-Hungarian Empire was characterized by the strong political stance of the Catholic Church, who supported the minority groups as a Habsburg loyal conservative stronghold against the emerging liberal ideology of German and Hungarian nationalists. The federal region of Burgenland, which became part of the Austrian Republic only in 1921, shows different patterns drawing on its heritage from the Hungarian part of the Austro-Hungarian Empire. There, little politicization for national parties combined with church-run schools and diverse local dialects into a 'village ethnos' that after 1921 could be integrated easily into the main Austrian political parties, the Social Democratics and Christian Socials. In contrast to the Carinthian Slovenes mobilized as part of the Partisan resistance, there was – with the exception of Jews and Roma – little National Socialist prosecution or nationally motivated resistance of ethnic minorities in Burgenland.

During the immediate post-war period 1945-1955, the political role of the Slovenes in Styria and Carinthia and the Burgenland Croats in the peace negotiations with the Allied powers led to their privileged minority status, defined by international and constitutional law in Art. 7 of the Austrian State Treaty. In 1976, the federal Minorities Act[4] enacted for the first time a unitary legal basis for all Austrian minority groups: not only the Slovenes and Burgenland Croats, but also the Viennese Czechs and the Hungarians were offered official status in a consulting mechanism to the federal government, the so-called Advisory Council ('Volksgruppenbeirat'). However, acceptance was low, as the Croats

---

[4]   "Volksgruppengesetz". *Fed. Law Gazette* No. 396/1976.

and Slovenes saw their constitutional rights reduced (Baumgartner 1995).

Nonetheless, the political escalation of the Carinthian conflict in the 1970s had given rise to new activism, resulting in the 1980s in political organization of minority interests outside the established party spectrum. A range of legal claims met a more liberal jurisdiction, leading from 1988 to the annulment of parts of the restrictive earlier legislation (Pernthaler & Ebensberger 2000; Rautz 2000; Pernthaler 2003). In 1989, the Slovenes participated in the governmental consultancy committee of the 'Volksgruppenbeirat'; in 1992 also the Viennese Hungarians were integrated into the Hungarian Council. In reaction to the separation of the Slovaks from the Czech Republic, the same year also saw the establishment of a separate 'Volksgruppenbeirat' for the Slovak nationality. In 1993, also the Roma and Sinti were included as a national minority group into the applicability of the federal Minorities Act of 1976. The 1990s saw also major regional improvements, particularly regarding bilingual schooling, official language use and bilingual topographic names in Burgenland. Despite two bomb attacks in bilingual villages in 1994/5, Burgenland's minorities remained mostly unconcerned by the nationality conflict in Carinthia. They profited from the improvements promoted by the Carinthian Slovenes at the federal level, and from the lessons drawn by the regional government of Burgenland for a more harmonious nationality regime (Baumgartner 1995).

Taking the geopolitical changes in Europe as an opportunity for regional development, the government of Burgenland aims to turn its border situation into a capital factor for regional development. Stemming partly from the time before the border opening in 1989 and before Austria's EU accession in 1995, a range of more or less formalized cross-border institutions has been established. Several bilateral cooperation agreements were initiated in the 1980s with Hungary on culture, with the independent Yugoslavian republics of Croatia and Slovenia on research, and some contacts with the then Soviet republic of Moldavia and with former Czechoslovakia. The exchange with Hungary became more institutionalized with the 'Österreichisch-Ungarische Raumordnungskommission' (ÖUROK) in 1985, followed by the 'Österreichisch-Ungarischer Regionalrat in 1992, and resulting in 1998 in the Euregio West-Nyugat Pannonia. Presently the most important cross-border project of Burgenland's government, the Euregio comprises the region of Burgenland and the neighboring Hungarian comitates of Gyor-Moson-Sopron and Vas. But it overlaps geographically with other preexisting transnational and transregional arrangements with different geographical and function ranges.[5] Thus, EU funded projects as well as public-private partnerships, semi-private development agencies, and inter-governmental coalitions serve as strategic instruments offering political, economic, and symbolic resources for regional development. While socio-economic and cultural gaps between the Hungarian and the Austrian side cannot be overcome quickly, the border region profits from large amounts of EU objective 1 funding channelled to Burgenland and from the market integration effects benefiting the Western parts of Hungary (Horvath & Müllner, 1992; Seger & Beluszky, 1993; Eger & Langer, 1996; Burgenländische Landesregierung *et al.*, 2000; Schimmel, 2001; Mayer, 2002).

The functional transformations within and across the borders with the kin-states provide new opportunities for collective action by the regional government as well as by non-governmental organizations. Rapid socio-economic modernization tendencies including economic restructurings, urbanization and cultural assimilation affect the bilingual areas as well as the whole region. Also cultural pluralization trends since the early 1980s have resulted in legal improvements, advertizing Burgenland as one of the most minority-friendly regions in Austria. These changes in the field of nationality politics, its cultural and territo-rial conditions, might all be associated with external changes associated with European integration. As a field highly sensitive for the sover-eignty of the member states, the national minority culture rarely ever becomes the direct focus of EU policies. It is only in the context of political-economic development policies, that cultural contents are pursued, e.g. in the framework of tourism development or employment qualification programmes funded by EU structural funds, and – to a very small budgetary extent – the promotion of language diversity. But the political-economic integration put forward by the EU contributes supra- and transnational patterns to the international environment characterized since the late 1980s and early 1990s by increasing ex-change across Austria's borders. European integration – in its broad sense – might thus offer a general opportunity structure and cultural frame motivating bottom-up change in the domestic policy field of nationality politics which is not directly affected by EU policies.

---

[5] Burgenland is a member in the Alps-Adria region to the West, in the Centrope project promoted by Vienna to the East, the ARGE Donauländer including the whole Danube region, the Assembly of European Regions and the 'Europäische Konferenz der Weinbauregionen', and in inner-Austrian cooperation with the regions of Vienna and Lower Austria in the Planungsgemeinschaft Ost and the Verkehrsverbund Ostre-gion.

III. Cultural Mobilization for Regional Development

Researching the political-economic functions and cultural meanings constituting the border, the following chapter focuses on the strategic re-orientation of the regional government policy from the national context toward an increasingly also supra- and transnational environment. How does the reorientation of the regional government's economic development policy from the national to a larger European space affect institutional changes in the field of minority policy? Do these cultural strategies also entail new structural opportunities for the region's national minorities? Or do they merely serve political-economic purposes through symbolic mobilization and representation? Analyzing the regional government's cultural policy in different political arenas at the regional, federal, and European level gives insight into the relationship between symbolic discourse and real political engagement for the nationalities in the region.

Since the beginning 1990s, the attitude of Burgenland's regional government toward the three autochtonous minority groups has changed importantly. The neglect of minority issues characterizing official visions of Burgenland since its post-war integration into the Austrian state slowly gave rise to a more tolerant and even positive attitude toward the region's multinational and multilingual heritage. This political turn also found expression in the governmental declaration by Regional Governor Hans Niessl (SP) at the beginning of his first coalition period in 2001:

> It is our objective to successfully strengthen our country – we want to ascend to the upper class of successful European regions. With this ambitious aim, I start my work as a regional governor together with the team of this regional government.
>
> (…)
>
> The multilinguality of our country serves us therefore as an additional start advantage because we can thus achieve easier access to the markets of Hungary and Croatia. In this sense the nationality groups also take an important bridge function in the economic realm.[6]

---

6   "Unser Ziel ist die erfolgreiche Stärkung unseres Landes – wir wollen in die obere Klasse der europäischen Erfolgsregionen aufsteigen. Mit diesem ehrgeizigen Anspruch trete ich als Landeshauptmann gemeinsam mit dem Team dieser Landesregierung an. (...) Die Mehrsprachigkeit unseres Landes kommt uns dabei noch als zusätzlicher Startvorteil zugute, weil wir damit leichteren Zugang zu den Märkten in Ungarn und Kroatien erhalten. Den Volksgruppen kommt in diesem Sinne auch eine wichtige Brückenfunktion auf der wirtschaftlichen Ebene zu". Regierungserklärung of Landeshauptmann Niessl, Burgenländischer Landtag, XVIII. Gesetzgebungsperiode – 2. Sitzung – Donnerstag, 1. Feber 2001, p. 33-40.

Also the new president of Burgenland's regional parliament, Walter Prior, formerly known as a hardliner for German assimilation amongst his SP party colleagues, joined the intercultural discourse at the opening of the legislation period in December 2000:

> The new regional parliament (…) is in its constitution a good example of the diversity of our country. It shows the broad spectrum of political orientations, which ultimately want to work each in their own way positively for the country. In the new regional parliament becomes expressed also the confessional diversity of our country. Last but not least, the constitution of the regional parliament documents also the language diversity, the richness of the country based on its nationalities. We thus continue a tradition of Burgenland appreciated widely beyond the borders and distinguished Burgenland above all by tolerance, mutual understanding and togetherness.[7]

While such intercultural commitments presented a new feature of official discourse characterizing the governmental period under Hans Niessl (SP) since 2000, the political struggles preparing this change already started under his Social Democratic predecessors. Already as early as in 1990, the Burgenland village of Kittsee offered the strategic location for a conference of nationality groups ('Volksgruppenkongreß') claiming the transformations in Eastern Europe as an opportunity for the realization of the rights of ethnic minorities for language and cultural diversity in a new Europe.[8] The main political yardstick constituted a study about the Croats in Burgenland, mandated by the Austrian Federal Chancellory upon the initiative of the Advisory Council and conducted in 1994 by a private market research institute (OGM 1994). At this occasion, Martin Ivancsics, then chairman of the Croatian Advisory Council and by 2000 also political secretary of the Regional Governor described the decrease of the Burgenland Croat language as 'identity problem not only for the Croats, but for the identity of Burgenland'.[9]

---

[7] "Der neue Landtag (...) ist in seiner Zusammensetzung ein gutes Spiegelbild der Vielfalt unseres Landes. Es zeigt sich die breite Palette politischer Strömungen, die letztendlich jede auf ihre Weise positiv für das Land arbeiten wollen. Im neuen Landtag wird auch die konfessionelle Vielfalt unseres Landes deutlich. Nicht zuletzt dokumentiert die Zusammensetzung des Landtages auch die sprachliche Vielfalt, den Reichtum des Landes durch seine Volksgruppen. Wir setzen damit eine weit über die Grenzen des Landes hoch geschätzte Tradition des Burgenlandes fort, die sich vor allem durch Toleranz, gegenseitiges Verständnis und durch das Miteinander auszeichnet". Ansprache des neugewählten Landtagspräsidenten, Walter Prior, Burgenländischer Landtag, XVIII. Gesetzgebungsperiode – 1. Sitzung – Donnerstag, 28. Dezember 2000, p. 8.

[8] "Wandel im Osten als Chance für die Volksgruppen". APA 30 Sept. 1990.

[9] "OGM: Situation burgenländischer Kroaten "alarmierend"". APA 15 Dec. 1994.

During the 1990s, several improvements for Burgenland's minorities became apparent: in 1993, the constitution of the Advisory Councils for the nationalities of Croatians, Hungarians, and Roma respectively had enabled the representatives nominated by the organizations in agreement with the Federal Chancellory a consultation status to the federal and – in fact rarely applicable – to the regional government on issues concerning the situation of the minority group. On this basis the Federal Chancellory increased the subsidies for the Austrian minorities from 0.36 million Euro since the enactment of the Minorities Act in 1977 up to around 3.8 million Euro yearly after 1995. Regarding the allocations for Croatian, Hungarian and Roma minorities, the minorities associated with Burgenland received around 1.8 million Euro in 2002.[10]

In 1994, a new federal Minorities School Act for Burgenland[11] replaced the regional legislation of 1937: it introduced the possibility to unregister from bilingual schooling in the autochtonous villages. At the same time, it complemented these territorial rights limited to the autochtonous territories by a personal right applicable in the whole region of Burgenland. It also extended the bilingual schooling offers from four years primary school to the full eight year period of compulsory education and reduced the numbers of students per class.[12] Accompanying public investments into bilingual education, culture, and schooling included also the establishment of a bilingual high-school in Oberwart and accompanying high increases of federal finance for schooling.[13]

At the regional level, the Kindergarden law of 1995 was the first major step also toward a regional legislation for the promotion of bilinguality. Yet, it was only by the year 2000 that the political situation in the region was ready for the symbolic act of installing bilingual topographic signs indicating the names of some autochtonous villages, already provided for in the Austrian State Treaty of 1955.

These various changes in the field of nationality politics were facilitated by several contextual factors: generational changes of the political elites leading from the 1980s to a pluralization of socio-political affilia-

---

[10] This estimate is probably a bit high because not all of the 1,196 million allocated to Croatians, 278,000 to Hungarians and 382,000 Euro to Roma (Rechnungshof 2004, p. 10) actually subsidized activities located in Burgenland or associated with the region; see overview of federal subsidies in Annex.

[11] "'Minderheitenschulgesetz für das Burgenland". *Fed. Law Gazette* No. 641/1994.

[12] HKDC & Landesschulrat, 2004, p. 112-113

[13] In 1996 the bilingual schooling subsidies for all of Austria amounted up to 90 million ATS in addition to 13 million ATS other minority project funding, making the Federal Ministry of Education and Culture the most important source for the promotion of bilinguality (Federal Government Report to Federal Parliament, 2002).

tions in the Austrian party system and the minority organizations;[14] the opening of the Iron Curtain from the end 1980s leading to symbolic reinterpretations of the region's border-situation from national periphery to European center;[15] Austria's EU accession leading to the symbolic re-evaluation of cultural diversity in the context of the EU's decision to accord Burgenland objective 1 status;[16] EU enlargement and increased economic competition leading to efforts for integrating the competition from Western Hungary into common institutions of cross-border cooperation.

EU-funds provided only an additional motivation for the regional government's earlier autonomous engagement to establish cross-border cooperation with Western Hungary. Mostly the EU funding guidelines and programme documents established economic and infrastructural policy priorities and little realm for cultural promotion. Compared actually to the overall amount of EU monies of 165.6 million Euro for the EFRE programming period 1996-2000 and the 271 million Euro for 2000-2006, the EU payments for projects coordinated by associations active in cultural minority or language promotion, namely a total of 1.935 million Euro under EFRE 1995-1999 and 149,164 Euro under INTERREG IIIA 2000-2006, seem quite small.[17]

Adding also the federal subsidies, which can be estimated in total between a third and a half of the yearly budget of around 3.8 million Euro from the Federal Chancellory and 7.4 million Euro from the Federal Ministry of Culture and Education, the cultural field attracted high increases of external funding to the region since the mid-1990s. The regional funding for minority culture is comparatively low and unstable, between 1.2 million Euro in the anniversary year of 2001 and under 100,000 Euro in 2002 out of a total regional budget averaging 884.5 million Euro.[18] 'Real' political action by the regional government in the area of minority policy focused less on the direct financial promotion of minority culture, but aimed to influence the federal policy to increase federal funding and infrastructural investments in the region. In the changing European context, the political discourse advertizing Burgenland as multicultural region contributed to attract these additional finances from the EU and from the federal state.

---

[14] Interviews 17, 22, 34.

[15] Interview 17.

[16] Interviews 8, 19.

[17] Wagner *et al.* 2003; statistics EFRE (07/2005) & INTERREG IIIA (04/2005), provided by Regional Management Burgenland.

[18] Burgenländischer Kulturbericht 2000; 2001; 2002; 2003.

Beyond these symbolic mobilizations for regional development, the regional government pursued few activities in federal or European minority politics. The leading politicians promoting Burgenland's cultural turn avoided any political confrontation with the national-populist governor of Carinthia, Jörg Haider. There are no representatives from Burgenland involved in the so-called 'consensus conference', which has been established by the Federal Chancellory as a mediation instrument for the nationality conflict in Carinthia.[19] Also there is little opportunity for minority associations to integrate their long established cross-border contacts and cultural knowhow into the emerging intergovernmental institutions governing the cross-border region.[20] The legal provision defining the whole region as bilingual schooling territory with mixed territorial and personal rights can be understood as a response to increased socio-economic mobility. Faced with unclear territorial definitions of the federal minority legislation, it countered a possible loss of financial claims to growing autochtonous and allochtonous Croatian, Hungarian, and Roma communities in the urban centers outside the region.

Thus, Burgenland's cultural turn provided symbolic and financial benefits for the region in competition with the growing political economy of the neighboring metropolitan center, Vienna. Just as the official neglect of the cultural heritage of the border region had served regional integration into the Austrian national economy, its recent rediscovery has contributed to the competitively advantaged position of the region in the larger European economy.

IV. National Minorities in a Changing Political Economy

The political-economic and social transformations in Europe might thus appear to contribute to an easy functional solution to the nationality problem of the region. The long neglect by the regional government might prove to a successful strategy of 'sitting out' the problem until it disappears by itself – either by assimilation or by changing external interest structures. But cultural belonging is deeply-rooted and – while not primordially defined – it cannot be transformed easily according to the respectively powerful political economic interest. The relations between majority and minority groups are based on long institutionalized paths guiding the political interactions within and between the national organizations. This poses the question how the different politi-

---

[19]   Interview 10, 17.

[20]   Interviews Göttel, 19, 32; the purely intergovernmental structures of the Euregio are provided for by Arts. 7, 8, 9 of the Framework Agreement about the Cooperation of the Euregio West-Nyugat Pannonia, see Burgenländische Landesregierung (2000).

cal actors in the field of minority politics interpret and respond to socio-economic re-territorialization processes and the resulting changes in regional policy. Do the various minority organizations adapt their cultural and territorial representations to the dominant regional consensus? The minority institutions vary according to their language or ethnicity, the socio-cultural or political-functional power basis of their organizations, their political party orientation, the legal status of their minority rights, their relationship with a kin-state, the citizenship status possibly defining the minority, and the link with their autochtonous territory. The diverse mobilization strategies of different minority groups illustrate how these diverse historic paths respond to changes in the institutional structures by different interpretations of cultural and territorial belonging.

The Burgenland Croats were polarized since the 1920s between Social Democratic promoters of national assimilation and Christian Conservative supporters of national segregation. During the Second Republic, this ideological polarization into two socio-political camps continued under the leadership of the Croatian Cultural Association of Burgenland 'HKD',[21] founded already in 1921 at the time of Burgenland's integration with Austria, and since the main organizational base of Croatian minority culture, promoting Christian-Conservative values and ethnic traditions of the rural communities; and on the Social-Democratic side, the socio-economically progressive Presidential Conference of Social Democratic representatives from Croatian and bilingual municipalities in Burgenland,[22] founded in 1978. Due to this link between ethnic and political organizations, Croatians came to work in all fields and for both political sides including important leadership functions in the regional institutions as well as in the federal parliament. Although there is no ethnic mandate in the regional parliament, its representatives have always included minority members. This functional inclusion actually implied little political representation of minority issues, attributing to the Burgenland Croatians a reputation as a 'silent minority'.[23]

Their silence came to an end during the 1970s when a new generation of activists claimed minority rights as fundamental democratic rights, a movement which by the 1980s gave rise to organizational

---

[21] HKD – Hrvatsko kulturno društvo u Gradišću / Kroatischer Kulturverein im Burgenland, Eisenstadt.

[22] Prezidij SPÖ-mandatarov iz hrvatskih i mišanojezicnih opcin u Gradišcu / Präsidium der SPÖ-Mandatare aus kroatischen und gemischtsprachigen Gemeinden im Burgenland, Eisenstadt.

[23] "Sprachenerhebung: trotz Minderheiten wenig Interesse im Burgenland". APA 11 Nov. 1976.

mobilization outside the established party spectrum. The Cultural Association 'KUGA'[24] in Großwarasdorf/Veliki Boristof, founded in 1982, aimed at critical, modern, non-folkloristic minority identification, cooperating also with parts of Viennese associations, particularly the Croatian Academics Club 'HAK',[25] a students association already established in 1948. These efforts resulted in a generational break with the party-linked minority organizations supported by traditional local elites who had turned into political representatives of mass regional organizations. Younger, educated, and urbanized elites, socialized student movements, pushed the demands for social reform and multi-cultural self-expression into the minority groups, and questioned the established power structures and their integration with the political establishment. The theoretical self-reflection and identification of cultural minorities with other socially marginalized groups gave rise in practice to a broad range of cultural activism outside the party-political field (Holzer 1993; Baumgartner 1996; 1999). As a result of this cultural and organizational fragmentation of minority mobilization, the political consultation instrument of the Advisory Council provided for by the federal Minorities Act of 1976 was not constituted until the 1990s. Yet, this generation's ideas of pluralism and interculturalism ultimately found entrance into the official governmental strategy by the year 2000.

For the Social Democratic side among the Croatian speaking population, the 1990s developed as a period of internal struggle over different cultural strategies of socio-economic progress: on one side, the established party elites pled for hesitant and gradual adjustments while they defended the old path of German assimilation as the only way the low-skilled working classes among the Burgenland Croatians could find employment, commute to Vienna, and thus integrate the peripheral region into the Austrian national economy.[26] Conversely, the opening of the borders with the neighboring states and Austria's EU accession provided an opportunity for the new generation of educated urban Croatian elites to reinterpret bilingualism and intercultural understanding as a means of socio-economic development. One of the first public expressions of the new intercultural proponents constituted a brochure, which provided an argumentation guideline for the promotion of bilingual teaching and education:

---

[24] KUGA – Kulturna zadruga / Kulturverein KUGA, Großwarasdorf.

[25] HAK – Hrvatski akademski klub / Kroatischer Akademikerklub, Wien-Eisenstadt.

[26] "[Walter] Prior: ORF-Sendungen für kroatische Volksgruppe zu begrüßen", OTS 7 May 1990; "Ortstafelfrage: SP-Mandatare setzen auf Prinzip der Freiwilligkeit", APA 28 Apr. 1994.

The first years in the EU have confirmed the general trend also for Burgenland's employment market: today the employees need a solid education and must be able to adapt to new developments. Increasingly also a further qualification gains importance – language knowledge! Who speaks two or more languages has definite start advantages in the competition for good jobs![27]

The minority organizations associated with the Social Democratic party or functionally linked with the regional government turned to the promotion of intercultural communication and bilingualism. Particularly, the Croatian Culture and Documentation Center 'HKDC', the School for Adult Education of the Burgenland Croats 'HNVŠ', the Association of Burgenland Croatian Pedagogues 'ZORA", and the Scientific Institute of the Burgenland Croats 'ZIGH'[28] contribute significantly to the development of teaching materials and schooling plans for bilingual language education in Burgenland.

Most socio-cultural associations associated with the conservative spectrum continued their engagement focusing on the promotion of Croatian culture as the representation of a separate ethnicity. The Croatian Cultural Association of Burgenland 'HKD' supported the intercultural turn of the regional institutions, but the decreasing use of Croatian as an everyday language undermined its political claims for ethnic representation. Therefore it focused its cultural-conservative engagement to local socio-cultural education activities within the autochtonous villages. As the ethnic-national conflict transformed into an issue of regional language and education policy, the conservative side now promoted Croatian language, music and craft as means for the conservation of traditional ethnic group belonging in local communities.[29] Despite – or possibly because of – commuting, urbanization, and subur-

---

[27]   "Die ersten Jahre in der EU bestätigen den allgemeinen Trend auch in der burgenländischen Arbeitswelt: die Arbeitnehmer/innen brauchen heute eine solide Ausbildung und müssen sich neuen Entwicklungen anpassen können. In verstärktem Maße kommt aber eine weitere Anforderung zur Geltung – Sprachkenntnisse! Wer zwei oder mehr Sprachen beherrscht, hat eindeutig Startvorteile im Wettbewerb um gute Jobs!", Ivancsics, M. 1998. "Warum nicht? Argumente für das zweisprachige Schulwesen / Zac ili Zasto ne? Argumenti za dvojezicno skolstvo". Eisenstadt / Zeljezno: HKDC – Kroatisches Kultur- und Dokumentationszentrum / Hrvatski kulturni i dokumentarni centar, 3rd ed., p. 6-7.

[28]   HKDC – Hrvatski kulturni i dokumentarni centar / Kroatisches Kultur- und Dokumentationszentrum, Eisenstadt; HNVŠ – Narodna visoka škola Gradišcanskih Hrvatov / Volkshochschule der Burgenländischen Kroaten, Eisenstadt. ZORA – Društvo Gradišcanskih pedagogov / Verein burgenländisch-kroatischer Pädagogen, Eisenstadt; ZIGH – Znanstveni institut Gradišcanskih Hrvatov / Wissenschaftliches Institut der Burgenländischen Kroaten, Eisenstadt.

[29]   Interview 16.

banization, the local associations experienced a revived interest in folkloristic leisure and entertainment activities amongst the younger generation (Holzer 1993).

The Croatian associations active outside the region of Burgenland, particularly the Burgenland Croatian Culture Association in Vienna 'HGKD' and the Croatian Academics Association 'HAK' as well as the Croatian Press Association 'HŠtD'[30] pursued a more pan-nationalist strategy. Addressing the Burgenland Croatian commuters and migrants in Vienna, these organizations extended their engagement beyond the autochotonous villages of origin to the capital city, the federal state as well as beyond the state-borders. Mutual interests in lobbying for bilingual Croatian schooling in Vienna as well as the common use of infrastructure available since the foundation in 1994 of the Burgenland Croatian Center 'CGH'[31] led to contacts with the allochtonous Croatian migrants' organizations. However, this mostly functional cooperation of the organizational elites contributed little to overcome mutual mistrust with regard to the nationalist mobilization in the Croatian kin-state (Bozic 1998). Opposing some migrants' claims to incorporate the Burgenland Croats within contemporary Croatian pan-nationalism, the autochtonous Croats stressed their loyalty to the Austrian state. They distinguished themselves as an old diaspora settled in the countries of the Austrian monarchy since the 17th century. Their historic homeland covered today's regions of Burgenland, Western Hungary, Slowakia, and Southern Moravia in todays' Czechia, from Southern Burgenland to Brno, from Vienna to Gyor, Marchfeld until Laa/Thaya, an area almost continuously inhabited by Croatians before 1921. The present works for language codification of a distinct Burgenland-Croatian dictionary that yet needs to include new words from Croatian standard-language are paradigmatic of the struggles to maintain the autochtonous identity separate and alive.[32]

The intercultural turn in the region of Burgenland is seen very critically by proponents, most of whom descend originally from conservative Croatian backgrounds, of an autochtonous national identity amongst the liberal educated elites. This skepticism is motivated by the fear that the opening-up of a weakened ethnic minority to intercultural cooperation risks, in fact, full integration or – in more radical terms –

---

[30] HGKD – Hrvatsko Gradišćansko kulturno društvo u Beču / Burgenländisch-Kroatischer Kulturverein in Wien; HAK – Hrvatski akademski klub / Kroatischer Akademikerklub, Wien-Eisenstadt; HŠtD – Hrvatsko štamparsko društvo / Kroatischer Presseverei, Eisenstadt.

[31] CGH – Gradišcansko-hrvatski Centar / Burgenländisch-kroatisches Zentrum, Vienna; Interview 9.

[32] Interview 12, 18.

assimilation, such as that developed under the former German strategy, that might take full effect even if disguised as intercultural dialogue. Moreover, the Viennese elites' suspicion against the intercultural turn in their homeland is also motivated by their exclusion from the regional compromise which mainly concerned the territorial interests of Burgenland.[33] Thus, the socio-political cleavage dividing the Croatian minority within Burgenland came to be covered under a territorially motivated institutional compromise that created new tensions between the Burgenland Croatians in the region and those outside it.

Although the opening of the borders resulted in a revaluation of the Hungarian language mostly among the majority population and business, this had little effect upon the Hungarian minority. The opening of the borders with Hungary initiated a symbolic revival of Hungarian heritage in Northern Burgenland, manifested in 1996 by a meeting of around 600 former inhabitants in the depopulated settlement Meierhof Albrechtsföld (Baumgartner 1999). In Southern Burgenland, the cultural activities carried by a few educated elites around the Hungarian Media and Information Center 'UMIZ'[34] promoted information exchange about Hungarian contemporary culture in a larger Pannonian region that crosses the state's borders. Unlike the Croatians, the Hungarians and their associations are weakly integrated into the regional political party system, and therefore they retrieve fewer political gains from the government's intercultural turn. Their stronger cultural and geographical dispersion among different localities has left the Hungarian minority even more weakened by socio-economic mobility and assimilation. Even more than the Croatians the Hungarian organizations suffer from a decreasing membership base and personal power struggles between the representatives.

These political divisions go back to the immigration waves of the 1950s and 1960s which strengthened the Hungarian populations in the urban centers outside Burgenland. The territorial cleavage between autochtonous organizations in Burgenland and allochtonous associations in Vienna is also a political cleavage defined by relationships with the Hungarian kin-state. The opening of the borders and improving bilateral relations between Austria and Hungary in the 1990s have strengthened the former Viennese dissidents numerically and politically who now re-established their relationship with the post-communist kin-state. The Burgenland organizations responded to reproaches for their

---

[33]   Interview 9, 12, 18, 25.

[34]   UMIZ/MMIK – Ungarisches Medien und Informationszentrum / Magyar Média és Információs Központ, Unterwart; Interview 24.

former Communist-friendly orientation by retreating to local activities dispersed amongst several villages.

Since the beginning 1990s, the mobilization and common representation of different Roma groups as one nationality group ('Volksgruppe') was supported by diverse contextual factors: the cultural or political engagement of some individuals in the group;[35] linguistic research and the standardization of the Burgenland Romani language; a general trend toward cultural pluralism and acknowledgement of diversity; the official acknowledgement of national-socialist crimes committed against Roma and compensation efforts by the Austrian state; the federal subsidizing structures necessitating the constitution of a common Advisory Council; as well as the policies of the Council of Europe and the OSCE.

In Burgenland, Roma associations have specifically organized in the district of Oberwart since 1989, namely in the 'Verein-Roma' and the 'Roma-Service'. Though they were born under the structure of a singular organization working for the promotion of education, the former focuses now on socio-economic integration projects for unemployed youth, while the latter concentrates on the development of teaching materials and bilingual literature. Moreover, the 'Kulturverein Österreichischer Roma', founded in 1991, is strongly anchored in a Burgenland Romani background but is situated in Vienna and more oriented in its political representation work toward the municipality and the federal state. 'Romano Centro', the other Viennese organization founded in 1991, understands itself as an allochtonous association of Roma with either Austrian or foreign citizenship, who immigrated since the 1960s (Baumgartner & Freund 2005).

As the degree of organization amongst the Roma is still very low, their political representatives pursue a highly inclusive strategy addressing allochtonous as well as autochtonous populations independent of their origin inside or outside Burgenland or Austria. Political organization and integration into the Austrian institutions is promoted by representatives who stress their personal autochtonous status and Austrian citizenship. Most political leaders associate themselves with the Burgenland Roma, a group long settled in the territory of today's Burgenland but whose population is recently decreasing relatively when compared to the Roma populations of mixed immigrant background in Vienna and other urban agglomerations. Regarding the education and development gaps to the organizations in neighboring Eastern European countries, the Austrian organizations see themselves as leading partners

---

[35]   See Stojka (1988): this autobiographic book is considered as one of the first public expressions of Roma culture in contemporary Austrian society.

in the emerging European cooperation for Roma promotion. Since 1993 the Karl-Franzens University in Graz works on the codification of the Romani language. The resulting Burgenland Romanes dictionary provides the Austrian organizations with a central role among all European groups in the further efforts to promote an international standard Romani language. Despite emerging European institutional cooperation and occasional allocations from the regional intercultural initiatives, the Roma associations gain most finances from federal minority subsidies and compensation funds.[36]

In sum, the promotion of multilinguality and interculturalism advertised by the regional government meets little open contestation by cultural minority associations. Yet, their different historic developments, organizational structures, legal situations, and territorial identifications motivate different interpretations of and strategic responses to the present socio-economic re-territorialization processes and the resulting political changes. Due to their close personal linkages with the political institutions of the region, the organizational structures of the Croatian minority are more directly affected by the emerging intercultural consensus than the Hungarians and Roma. While some Roma have recently received increased acknowledgement from majority political institutions, the Hungarians seem to instead become more frustrated about their future as an ethnicity. Similarly, the Burgenland Croatians in Vienna see fewer benefits as an ethnic minority left out of the intercultural strategy established by the Croatian elites in the region of Burgenland.

V. Political Winners and Losers of Multinational Regionalism

Regarding these diverse cultural paths of the minority groups associated with Burgenland's territory, the changes of the government's cultural policy might have advantaged some and disadvantaged others. Can we then speak of an emerging dominant vision that strengthened the power of the regional institutions not only in the political-economic sphere but also in the cultural field? Which social and political actors in the field of minority politics win or lose from cultural mobilization for the new regionalist policy? The following section analyzes the resulting structural changes to the public institutions of nationality politics in Burgenland at the input-side as well as the output-side of regional cultural policy. Inclusion and exclusion in the dominant public vision of territorial culture constitutes varying boundaries between the public and

---

[36] Interviews 2, 28.

the private sphere of minority promotion, the political and the social sphere of minority activism.

The emerging intercultural development vision strengthened the power of regional institutions in the field of minority policy mainly in that it enabled different regional actors to overcome historic, political, and cultural divides and implement a common action at the federal level. Starting from the mid-1980s, this cultural turn was facilitated by the retirement of several leading figures, so that from the early 1990s, parts of the younger generation proceeded into leading positions in minority rights organizations and political parties. It was on the basis of this generational change that a political compromise between the Croatian elites of both grand regional parties could be achieved. The infiltration of regional political institutions with Croatian minority elites, historically responsible for their nationalist polarization, enabled a political party compromise as a basis for intercultural regional development.

Only through such cooperation was it possible to constitute the Advisory Council for the Croatian nationality in 1993,[37] seventeen years after the legal provision by the Minorities Act of 1976. According to the federal provisions,[38] the Advisory Council was to be constituted by two chambers, one speaking for the political parties and the Catholic Church, the other one for the non-governmental organizations representing the minority. However, this requirement contradicted the structures of the Croatian organizations constituted by two large minority representative camps: the Presidential Conference of SP Mayors and political representatives in bilingual Croatian communities, founded in 1978; and the Croatian Cultural Association of Burgenland, founded in 1921 at the time of Burgenland's integration with Austria. The former gained its representative power from the political functions of its members as representatives of a political party (the SP) and would thus belong to the first chamber. The latter was only ideologically linked with the Austrian People's Party (VP) but as an association with a broad social membership base it would belong to the NGO chamber. This functional opposition of organizational structures enhanced the cultural-political deadlock which inhibited the constitution of the Advisory Council in the Austrian tradition of party corporatism based on elite cooperation and proportional representation of the two big political parties. The functional constellations for the party compromise were improved only from 1986 by the foundation of a party-based representation structure also on the

---

[37] "Aufgaben des Beirates". *Kurier* 23 Nov. 1993
[38] "Ordinance of the Federal Government governing the Advisory Councils for National Minorities". *Fed. Law Gazette* No. 38/1977.

conservative side. As the counterpart to the SP-based Presidential Conference, the Working Group of Croatian Municipal Politicians in Burgenland convened all VP-Mayors and political representatives of Burgenland's bilingual Croatian villages.[39] Now the necessary functional-political balance was provided to establish the Advisory Council with de-facto eight seats for the Social Democratic party and six of the conservative political side.[40]

Yet, the Advisory Council remained divided by the years-long discussion about the issues of topographic signs, which struggled regional SP elites.[41] It was only in July 2000, following a political change in the federal government, that the new, conservative Federal Chancellor Wolfgang Schuessel took the symbolic step to install the bilingual topographic signs in 47 villages of Burgenland. The motivations for this federal decision were interpreted differently: the political achievements for the conservative side were claimed and actually enforced by an ordinance enacted by SP Chancellor Viktor Klima during the last days of his government.[42] In the context of the EU sanctions against the extreme right-populist Freedom Party in the new federal coalition government, this symbolic act was useful as proof of the conservative government's respect for human rights.[43] Ultimately, it was claimed a victory for all political sides so that the cultural struggles dividing the region, and particularly the governing Social Democratic party, during the 1990s were terminated.[44]

[39] DZ – "Djelatna zajednica hrvatskih politicarov u Gradišcu /Arbeitsgemeinschaft kroatischer Kommunalpolitiker im Burgenland", *Kroatisch Geresdorf*; Interview 29.

[40] Of 24 seats in total, five each were allocated to the political parties SPÖ and ÖVP, two to the Catholic Church in the 'party curia'; still, in the 'party-independent curia' four seats were constituted by party-close minority organizations and only the remaining eight were taken by actually independent cultural associations. http://www.hkd.at/iinfode.htm (June 2005).

[41] "Burgenländische Seele wird zu den zweisprachigen Ortstafeln untersucht". *Kurier* 31 May 1994; "Ortstafelfrage: SP-Mandatare setzen auf Prinzip der Freiwilligkeit". APA 28 April 1994; "Zweisprachige Ortstafeln im Burgenland: Verfassungsklage droht". APA 13 April 1994; "SPÖ verzögert die Aufstellung von gemischten Ortstafeln im Burgenland". *Der Standard* 25 Dec. 1994; "Abstimmung über Ortstafeln". *Kurier* 8 Feb. 1994; "Durchbruch bei zweisprachigen Ortstafeln für das Burgenland". APA 18 Nov. 1993; "'Ortstafelstreit' auf burgenländisch". *Kurier* 6 Nov. 1993; "SPÖ verzögert die Aufstellung von gemischten Ortstafeln im Burgenland". *Der Standard* 25 Feb. 1994; "Erste Zweisprachige Ortstafel im Burgenland". APA 13 July 2000.

[42] "Topographical Ordinance for the Burgenland". *Fed. Law Gazette* Vol. II No. 170/2000

[43] Interviews 17, 29.

[44] In 2000, the newly elected regional governor Hans Niessl appointed Martin Ivancsics, the chairman of the Croatian Advisory Council and of the organization HKDC,

Thus capturing the Croatian minority institutions, the SP-governed regional government gained not only federal minority subsidies for the region but also informal decision-making power over their distribution. The Croatian Council prides itself on being the most efficient and consensus-oriented of all the national Advisory Councils because the party representatives normally coordinate their positions in advance. At the meetings there is then little realm for discussion or negotiation before the NGO mandatories are asked to vote on the proposals of the politicians.[45] Lacking more formal regional competences over minority issues and facing a very passive federal state, the smooth functioning of the Croatian minority institutions has provided the regional government of Burgenland not only a symbolic leadership role but also informal legislative power in the federal management of the region's minorities.[46]

In return, the regional government's official acknowledgement of cultural diversity brought a symbolic re-evaluation, increased political attention, infrastructural and financial improvements for all minority groups in the region. But these general improvements on the output-side are opposed to very unequal developments favoring the Croatian elites at the input-side of regional politics. To justify the intercultural turn the official discourse referred to economic necessities related to the opening of borders. But in fact, neighboring Hungary is a much more important economic partner than Croatia, which is not even an EU accession candidate. The growing interest amongst the German-speaking majority population in Hungarian language is also reflected by the bilingual schooling statistics and the increase of bilingual schools established since 1994 outside the autochtonous villages.[47] Yet, the increased

---

as his political secretary. Also the SP-hardliner Walter Prior declared his commitment to intercultural dialogue in his inauguration speech as a parliamentary president; see: Walter, P. "Ein Burgenlandkroate als Landtagspräsident". APA 28 Dec. 2000.

[45] Interviews 16, 17.

[46] As the Federal Court of Audit stated in its 2004 review, the Advisory Councils' decisions about the distribution of minority subsidizing are normally accepted and as such, they are implemented by the Federal Chancellory:

"10.1. Das BKA entsprach bei den Entscheidungen über die einzelnen Förderungen in vollem Umfang den Empfehlungen der jeweiligen Volkgruppenbeiräte.

10.2. Somit überließ das BKA de facto den Beiräten die Entscheidung über die Förderungswürdigkeit der Anträge. Es fehlten jedoch spezifische Förderungsrichtlinien, um die Praxis der Volksgruppenförderung ingesamt transparenter zu machen" (Rechnungshof 2004, p.11).

[47] During the schooling year of 2004/5, the numbers of students attending bilingual schools during nine years obligational schooling (four years primary school, four years secondary school) amounted 1,701 for Croatian teaching (1,435 primary, 266 secondary school; as well as 286 in highschool), 1,408 for Hungarian teaching

interest in the Hungarian economy and language was not transformed into a political-institutional advantage for the Hungarian organizations. Due to their strong links with regional political institutions, only the Croatian elites were in a position to gain symbolic leverage from the opening of Eastern Europe for 'Croatian as a Slavic language'.[48]

Most of the recent intercultural activities in the region are actually managed by representatives of the Croatian minority, often with close links to the political parties, particularly the governing Social Democracts. The beginning of the 1990s therefore saw increasing activities by some associations close to the SPÖ, namely the above mentioned HKDC, the HNVŠ, and ZORA. Moreover, the Scientific Institute of the Burgenland Croats ZIGH was founded in 1994. Traditional ethnic organizations such as HKD, HCKD or HAK all drew on a large social membership base from amongst the minority population, the voluntary work of their members, and the federal subsidizing of one long-served employee. But the newly emerging institutional infrastructure was organized around a few professional employees with a project-oriented work style that lacked a membership basis as well as basic federal subsidizing.[49] Thus, the politically mandated intercultural turn of the region allowed cultural minority politics back on the political agenda as an uncontroversial issue of semi-privatized development management. Historically, the personal links between Croatian minority organizations and regional political parties were responsible not only for the nationalist polarization of the minority but also for the exclusion of minority claims from the agenda of Burgenland's post-war consociational institutions. Now, the renewed de-politicization of minority culture as part of regional development undermined the political claims of the established structures of ethnic representation, particularly the Burgenland Croatian Cultural Association as the largest member-based minority organization in the region.[50]

This cultural de-politicization was also accompanied by an organizational politicization which strengthened the legitimacy of the political parties and government-related institutions to act on behalf of minority culture.[51] This structural politicization mainly favored the Social Democratic party which extended its representative power into the field of

---

(886 primary, 522 secondary, as well as 224 in highschool), and 24 in Romanes primary schools; see statistics in HKDC & Landesschulrat (2004).

[48] Interview 17.

[49] "Wissenschaftliches Institut der Burgenland-Kroaten gegründet". APA 31 Jan. 1994; Interviews 12, 18, 19.

[50] Interviews 16, 29.

[51] Interview 17.

minority politics. Also the Green Party, whose originally countercultural activism was now turned into a core vision of regional development, won a second seat for one of the founders of the intercultural grass-roots initiative KUGA in the regional elections of 2000. The Christian conservative Peoples Party (VP), though participating as the grand oppositional party in the regional intercultural compromise, lost ethnic conservativism as an issue of political mobilization. While being able to show financial gains for the ethnic activities, the party elites suffered from a political divide with the conservative Croatian cultural association and alienated large parts of the ethnically conscious electorate. The radical right Freedom Party also found little political ground for nationalist contestation given the dominant regional vision of intercultural economic development.[52] This trend in the field of minority policy was also reflected in the 1996 regional elections, the results of which marked this year as a turning point towards increasing electoral support for the Social Democrats and for the Green Party. These election results also showed a strong decline in support for the radical right-nationalist Freedom Party, and stabilization for the Christian-conservative Peoples Party.[53] While the changes in the field of minority policy were certainly not the main causes for the changing electoral trend, the concurrent cultural changes were certainly not opposed to these political tendencies.

VI. National Mobilization as Multi-level European Politics

Regarding Burgenland's intercultural response to European integration, the question must be posed whether those ethnically oriented actors left out of the new multilingual policy can make Europe their new political home. Is the much advertised 'Europe of the Regions' also a 'Europe of Cultures' in that it provides opportunities for national mobilization beyond the borders of the state? Do those who are excluded or losing from regional development turn to another territorial level of government or do they remain in the region to contest the regional consensus? As regional development policies seem to develop more autonomous strategies, these multi-level politics might also cause spill-over effects upon the territorial organization in the field of minority politics. The intergovernmental and transnational aspects of European

---

[52] Interviews 15, 19, 22.

[53] The regional election results were for the SPÖ 44.45% in 1996, 46.55% in 2000, 52.23% in 2005, for the Greens 2.49% in 1996, 5.49 % in 2000, 5.2% in 2005, for the F 14.55% in 1996, 12.63% in 2000, 5.76% in 2005, and for the ÖVP 36.06% in 1996, 35.33% in 2000; 36.34% in 2005; for graphical illustration of regional election results since 1945 see Annex.

integration in the national minority field will be analyzed by identifying the cooperation strategies of different minority actors and their main political arenas.

The Austrian federal state remains the main governmental level with formal competences in the field of minority policy based on constitutional and simple law.[54] Since the post-war peace settlements, the federal minority protection policy had been embedded in a changing international context. In the international post-war order, European kin-states were attributed an important protection role for national minorities in the neighboring states. The role of bilateral treaties is being reconsidered in the context of post-communist transformation and European integration. European integration might have a certain symbolic effect at the intergovernmental level, enabling EU members such as Hungary to take a more demanding stance toward the Austrian state to provide adequate minority protection. This intergovernmental pressure – together with the symbolic effects of EU conditionality and the Council of Europe conventions, and Austria's recent membership in the EU – provided the context for the federal government's new minority policy during the mid-1990s. But these pressures were mostly symbolic, mediated by the federal state, as illustrated by the fact that the optional clauses foreseen by the Council of Europe conventions obliged the Austrian state to no more than its domestic status-quo.

The international transformations fall short also of any relevant increase of direct influence of the kin-state upon the minority organizations or the regional government. The Austrian government declines to acknowledge Slovenia and Croatia as successor states of Yugoslavia with regard to its protection power as kin-state anchored in the Austrian State Treaty of 1955.[55] This shift from international obligations toward a more sovereign policy did not imply a weakening of Austria's constitutional obligations toward these minorities anchored in domestic law. At least in the case of non-EU member Croatia this did not provoke any contestations from the side of the kin-state or the minority organizations, who – for historic reasons and present political-economic difficulties – do not maintain any close contacts with their kin-state. Hungary, a politically and economically better situated EU-member, affords political and financial support for its expatriate organizations in Austria.[56]

---

[54] The only regional innovation is the Kindergartengesetz of 1995 which introduced the possibility of bilingual education in the preschooling age, and a decision 2005 by the regional parliament demanding federal finance compensating the additional public investments necessitated by the federal schooling legislation of 1994.

[55] "Laibach will Österreichs Slowenen mit Minderheitenvertrag schützen". *Die Presse*, 24 Feb. 1995.

[56] Interviews 4, 24.

Yet, the Hungarian minority in Burgenland is weaker than the Croatians, whereas the Roma received high increases in federal funding without the support of any kin-state. This comparison illustrates the closer relationship minority organizations have with domestic institutions than they do with kin-states, even in a context of border openings and EU-enlargement.

Also the EU as a supranational European policy framework exerts little direct influence upon the minority protection regime. Out of 307,965 Euro overall project budgets coordinated by May 2005 by cultural minority associations under INTERREG IIIA (2000-2006), only one project of 8,300 Euro was not directed by Croatians (it is run by Hungarians).[57] While the regional government funding for cultural minority activities seems to be more balanced, the Croatian-run organizations were the main carriers of EU funded projects. Unlike the federal subsidies for ethnic minority associations, EU structural fund guidelines do not permit the subsidization of cultural activities for their own sake. Therefore, promotion of intercultural heritage and multilingualism was subsumed under the territorial economic objectives guiding EU regional policy. The Croatian elites, due to their close links with the regional government institutions, shared this territorial vision and had access to the necessary political and infrastructural resources for such administration-intensive EU-projects. EU structural funding is mainly governed by the regional political executive, which not only drafts the programming document but also has the ultimate decision-making authority concerning project allocation. The semi-governmental regional management agency serves merely as a technocratic implementation arm, whereas the real decisions seem to be taken according to party-political interests in the regional Governor's office. The small chance for outsiders to win funding in this party-dominated system is exacerbated by the limited personnel capacities of mostly voluntary minority work, and the labor-intensive administration of EU projects. Thus, there is little bottom-up engagement of the longer established minority organizations for EU funding.[58]

Only the new intercultural grass-roots initiatives (KUGA, OHO, Europahaus)[59] could establish themselves through EU-funding. Being neither eligible for federal minority subsidies nor allowed to nominate any members to the Advisory Council, associations that were originally considered to be "a provocation to the political establishment", saw EU funding as their only financing opportunity. However, they were suc-

---

[57] 'EFRE, RMB Programm-Management'.

[58] Interviews 16, 26.

[59] OHO – Offenes Haus Oberwart; Interviews 19, 27, 31.

cessful only in those cultural programmes administered directly by the EU Commission. Only later, in the context of EU accession, was their intercultural engagement rewarded by their symbolic inclusion into the regional cultural institutions.[60] KUGA and OHO received high EU subsidies dedicated, according to the guidelines of objective 1, mainly to infrastructural investments. Lacking sufficient complementary funding for labor cost to manage and upkeep their new buildings, these countercultural initiatives are now shaken by financial and identity crises. Conversely, one initiative continuing its controversial programming, which contested and discussed the meaning of Europe, was excluded from regionally administered EU funds since the beginning. This association still maintains its controversial activities on a small level through direct funding from the EU Commission, but it was forced to slightly de-politicize its focus and redirect it away from the region of Burgenland, towards an international search for cooperation partners.

Generally, EU structural funds provided little additional input to initiate or promote cross-border cooperation among minority organizations. Mostly cross-border contacts existed already before the fall of the Iron Curtain. They were then revived in a rather unsystematic way, and sometimes facilitated by EU-funding. INTERREG or Comenius funds were mostly allocated to governmental institutions such as schools or municipalities, mainly in support of the cross-border activities of the regional government. Representatives of cultural associations saw the main advantage of European integration in the opening of borders, which facilitated existing cross-border contacts but will only take full effect with the implementation of Schengen-agreements from 2007.[61]

The cultural associations' various cultural exchanges, international contacts and cross-border activities rarely included any systematic political activities at the transnational or European level. They mostly stressed the irrelevance of their various external activities relative to their domestic ties with the Austrian state in terms of financial resources and loyal citizenship. The Viennese associations, particularly the Croatian Academics Club, which initiates and organizes a yearly Croatian youth festival in changing locations of the diaspora across Europe have

---

[60] It was particularly following the symbolic occasion of the EU Commissioner's visit to the region which included also a concert at the KUGA in Grosswarasdorf that Burgenland's objective 1 status was decided and the counter-cultural initiatives received funding; see: "Burgenland becircte EU-Regionalkommissar". APA 26 Nov. 1993; "EU-Kulturattaches besuchen Burgenland". APA 12 Oct. 1994; "EU ist burgenländische Kulturszene Millionen wert". *Kurier* 11 Oct. 1996; "Millionen für Alternativkultur". *Kurier* 25 March 1997.

[61] Interviews 21, 12, 23; see also Horvath & Müllner (1992); Zuckerstätter-Semela (2001); Baumgartner *et al.* (2002).

been the most active at the European level. Moreover, some of the leading members are active in the European Youth Foundation, an off-spring of the association more conservative FUEV for 20-30 years, where the president of the Burgenland Croatian cultural association is one of the board members. Although these two ethnic organizations are the ones losing most from Burgenland's intercultural turn, none of these transnational associations nor any of the supranational European bodies seem to provide a feasible alternative to the domestic state institutions.[62]

A more successful ethnic bridge for public mobilization across the state-borders seems to be the media. For about 25 years, the Croatian weekly newspaper 'Hrvatske Novine' has served a pan-national market with its 3400 copies and 15,000 readers, one third of them abroad, mainly in Hungary.[63] In 1989, Austria's public broadcasting corporation ORF introduced specific regional programs for national minorities, including a weekly TV programme, multilingual radio news and an internet news website.[64] In 1998, the privatization of radio frequencies enabled the establishment of a private bilingual radio programme called MORA. After a few years, however, a transfer of frequency rights that turned the non-profit dedication into a commercial purpose put an end to this grass-roots initiative around the KUGA association. Since then, the underrepresentation of multilingual media concerns a major discontent,[65] especially regarding the insufficient Hungarian programmes on public TV and radio. Only the internet page of the UMIZ provides extensive and detailed information about the region called 'Wart', which surrounds the village of Unterwart within and beyond Austria's borders with Hungary. However, run by its highly engaged organizer as a 'one-man-show', some suspect, its intellectual approach might fulfill the information demands of external elites rather instead of attracting interest from the regional population.[66] On the contrary, the Croatian webpage 'Cyberkrowodn' fulfills a more bottom-up demand for participation and popular culture. Emerging around a village restaurant in Unterpullendorf / Dolnja Pulja, it invites the users' contributions to go beyond a 'folkloristic traditional corset' represented by 'singing, jumping, and praying' and a high culture carried by 'academics, politicians, and priests'.[67] Thus, the various media initiatives aim to reflect a diverse contemporary spectrum of everyday minority culture beyond language

---

[62] Interviews 16, 9, 25.

[63] Interview 12.

[64] www.volksgruppen.orf.at; Federal Chancellory (2000).

[65] Interviews 18, 19.

[66] www.umiz.at; Interviews 24, 20, 25.

[67] www.cyberkrowodn.at.

and state boundaries. Yet, these initiatives carried by a few active individuals and groups can only cover people's private and leisure activities and can hardly extend the use of minority languages into the public sphere.

While the region of Burgenland has benefited economically, politically and symbolically within federal minority politics, it has abstained from using its power beyond its immediate regional interests. The regional policy for intercultural development focuses mainly on cultural promotion within the region and shows little external engagement for minority protection either within the federal or the European contexts or for 'expatriate' minority members. The federal Minorities School Act of 1994 complemented the constitutional territorial rights of some autochtonous villages by including personal rights applicable only in the region of Burgenland. These rights took account of increasing social mobility within Burgenland but not of increasing commuter and migration flows beyond the borders of the region or the state. Specifically, the growing Croatian communities in the neighboring metropolitan center and capital city Vienna feel left out of the regional intercultural consensus dominating the federal policy toward the Croatian minority.

The metropolitan center and capital city Vienna emerged as an alternative territorial base for all three Burgenland minorities. Responding to increasing immigration, the Viennese government recently stressed the integration of allochtonous minorities.[68] Due to a lack of constitutional rights for autochtonous territorial minorities, the Viennese Croatians' are more disposed towards cooperation with allochtonous Croatians. But this pan-national opening of this minority population was inhibited by the last census of 2001 which expressedly distinguished between Burgenland Croatian and the standard Croatian language. Conversely, this distinction was not made for the Roma or the Hungarian minorities, both autochtonous in Burgenland with increasing autochtonous and allochtonous populations in Vienna. This division served to maintain the existing federal minority regime distinguishing between autochtonous and allochtonous nationalities and it maintained the financial and political power of the Burgenland's political elites in Croatian minority politics. The intercultural development strategy served the regional interests of Croatian political elites in federal minority politics, but it offered little support to the Hungarian interests in the region.[69]

For their part, the Hungarian and Roma organizations could continuously rely on the federal government for financial support. The amount and distribution of the federal subsidies for these nationalities has

---

[68] Interview 5, 7, 9, 11, 30.
[69] Interviews 4, 20.

remained relatively stable since the sharp increase in 1995, following the initial constitution of the Advisory Councils. Compared to their relative size, the Slovene minority (around 1.2 million Euro yearly) and the Roma (more than 390,000 Euro) receive disproportionately high funding. The Hungarians' allotment (around 280,000 Euros) is extremely low, the Slovakians' (40,000 Euros) is slightly low, and the Croatians (1.2 million Euro) seems relatively fair (Rechnungshof 2004). The imbalance between the distribution of funding and the population numbers among the minorities does not reflect the increased influence of the Croatians in regional politics.

Yet, the federal minority regime finds itself in a growing crisis as its focus on collective and territorial rights tends to conserve the carrier organizations' structures without regard for demographic developments. Against territorial and cultural challenges from immigration and social mobility, the Federal Chancellory has stressed the impossibility of quantifying national belonging and the importance of organizational activities and territorialized structures for the maintenance and existence of the minority group. Following a slightly more active stance in the context of Austria's European integration policy during the 1990s, since around 2002 the right-conservative government has once again retreated to passivity, characterized by timely delays of payments and parliamentary reports. It has lacked criteria and evaluation of achievements of the legal objectives of 'conservation and guarantee of existing status of the nationality group'. It is also characterized by a reduction of personnel in the department responsible for national minority affairs of the Federal Chancelory's Office and a de-facto transfer of decision-making on subsidy applications to the Advisory Councils (Rechnungshof 2004). While a territorial and ethnic separation seems to be the intentional policy of the federal government, the emerging inconsistencies provide opportunities for new actors, thus deepening the regional asymmetry of the Austrian minority regime.

VII. Conclusion

The government of Burgenland seized the opening of Europe's borders for a development strategy that mobilized its border identity for a reorientation from the margins of Austria's national economy into the centre of the European market. This economic policy explicitly acknowledged the region's multi-national heritage, so far largely neglected by the regional government with political representation limited to cultural minority organizations at the level of the federal state. The regional development strategy resulted in a strategic turn of cultural policy from national assimilation to a post-national idea of an intercultural border region. While facilitated by the symbolic importance of borders in the context of European integration, the structural incentives for these changes in the field of minority policy came from the federal state.

The regional government's intercultural-bilingual turn is based mostly on a political reorientation of Croatian party elites responding to the socio-economic challenges from the opening of the border. The cultural organizations respond to the changing border context by adapting the territorial identifications associated with ethnicity to their various strategic interests. While rarely challenging the regional government's engagement in the cultural field, the minority associations tend to continue their separate ethnic paths within or beyond the state borders.

Their close links with the political party spectrum allowed the Croatian organizations to overcome their internal polarization between national integrationism and ethnic seperationism. The formerly excluded field of nationality politics could thus be included into the regional institutions. The political acknowledgement of the multinational heritage was motivated by the regional political elites' interest in a collective economic development effort. But beyond symbolic language this favored mostly the construction of a parallel organizational infrastructure controlled by Croatian government executives. Their intercultural cross-border ideology is little reflected by any structural integration of Hungarian or Roma representatives in the organizational leadership. The top-down politicization of an explicitly regional minority culture also tends to exclude minority representatives outside the regional boundaries of Burgenland. Prioritizing ethnic separation to an intercultural consensus that might imply yet another assimilation threat, there is an alternative tendency to pannationalist notions of ethnicity. While symbolically referring to the opening of borders, the government's intercultural strategy aims mainly at overcoming ethnic boundaries within the region. It thus takes into account the deepening of territorial and ethnic divisions in the minority field.

European integration gives rise to little direct political impact in the national minority field because culture remains mostly a domestic policy field. Yet, there are important challenges to the federal minority protecttion system, mostly stemming from socio-economic pressures, territorial mobility, and cultural mobilization in the context of changing border regimes. Burgenland's intercultural mobilization both responds to and poses a challenge to the federal minority protection regime. Their political party integration turn the Burgenland Croatians from Austria's silent minority into the best organized players. Providing the regional government direct access to federal minority policy, additional federal finances could be secured for the development of the region. But the cultural exchange programmes promoted by EU-cross-border funds are mostly run by governmental organizations and the traditional minority organizations profit little from EU funding. Despite their long-established cross-border contacts at the grassroots level, the minority associations are not included into the official cross-border region. They also find little alternative opportunities for political coordination at the European level.

These findings support the assumption that the transformation of border regimes in the context of European integration gives rise to changes in the cultural field due to a diversification of sub-national policy choices. However, minority groups mobilize in response to different political opportunities emerging either from functional effects from increased cross-border mobility and market integration, from EU policies supplying structural funding for cross-border cooperation, or from the changing symbolic meanings of border and transnational identities. Beyond extending the geographical realm of minority activities, the demographic transformations resulting from increased cross-border mobility also shift the cultural focus from rural to urban areas. Different strategic choices motivating the various actors have stressed the distinction between the cultural and territorial mobilization of ethnicity. The varying strategies of the Hungarian, Croatian, and Roma representatives also highlight the diversity of cultural paths despite their common territorial interest in regional development. Following the path of Austrian consociationalism, the political parties and the federal state ultimately remained the most important actors in the minority field. Marking the cultural and territorial boundaries of political institutions, state borders remain important gateways that limit minority politics and control the Europeanization of the domestic policy field.

References

Bauböck, R., Baumgartner, G., Perchinig, B. and Pinter, K. eds. 1988 ... *und raus bist du! Ethnische Minderheiten in der Politik*. Wien: Verlag für Gesellschaftskritik.

Baumgartner, G. ed. 1995. *6 x Österreich: Geschichte und aktuelle Situation der Volksgruppen*. Klagenfurt, Celovec: Drava Verlag.

Baumgartner, G. 1999. Sprachgruppen und Mehrsprachigkeit im Burgenland (http://www.ned.univie.ac.at/CMS/Brochüren/Musik_Sprache_Identität/Sprachgruppen_und_Mehrsprachigkeit_in_Burgenland/; Feb 2005)

Baumgartner, G., Kovacs, E. and Vari, A. 2002. *Entfernte Nachbarn: Janossomorja und Andau 1990-2000*. Budapest: Laszlo Teleki Stiftung.

Baumgartner, G. and Freund, F. 2004. *Die Burgenland Roma 1945-2000: Eine Darstellung der Volksgruppe auf der Basis archivalischer und statistischer Quellen*. Eisenstadt: Burgenländisches Landesarchiv.

Baumgartner, G. and Freund, F. 2005. *Roma-Politik in Oesterreich*. Kulturverein Österreichischer Roma & Fraktion der Sozialdemokratischen Partei Europas in Europäischen Parlament.

Bozic, S. 1998. *Immigranten und Integration im Zusammenhang mehrschichtiger ethnischer Beziehungen: Am Fall der Kroaten in Wien* (Forschungsprogramm 'Grenzenloses Österreich'; Schriftenreihe des Instituts für Soziologie, 37), Wien: Bundesministerium f Wissenschaft und Verkehr; Universität Wien, Grund- und Integrativwissenschaftliche Fakultät.

Burgenländische Landesregierung – Stabstelle Europabüro und Statistik & Ungarisches Statistisches Zentralamt, Komitatsdirektion Györ-Moson-Sopron, Vas, Zala. 2000. *Euregio in Zahlen: Eisenstadt, Györ, Szombathely, Zlägerszeg.*

Böckler, S. ed. 2004. *Minderheiten und grenzüberschreitende Zusammenarbeit im Alpen-Adria-Raum*, Trient: Arbeitsgemeinschaft Alpen-Adria; Autonome Region Trentino-Südtirol.

Eger, G. and Langer, J. eds. 1996. *Border, region and ethnicity in Central Europe: results of an international comparative research*, Klagenfurt: Norea.

Federal Government. 2002. Bericht der Bundesregierung gemäß § 9 Abs.7 des *Volksgruppengesetzes über dies Volksgruppenförderung in den Jahren 1997-2001 mit Anhang für das Jahr 1996* (III-96 der Beilagen zu den Stenographischen Protokollen des Nationalrates, XXII. Gesetzgebungsperiode)

Gmeiner, E.M. 1999. *EU Structural Policy and its implementation in Burgenland during the programming period 1995-1999*, Wien, Wirtschaftsuniersität.

Greif, F. 1993. *Regionalpolitik an gemeinsamer Grenze: Das Beispiel Österreich-Ungarn* (Schriftenreihe der Bundesanstalt für Agrarwirtschaft, 73)

Haslinger, P. 1999., ed. *Grenze im Kopf: Beiträge zur Geschichte der Grenze in Ostmitteleuropa*, Frankfurt am Main, et al.: Peter Lang.

HKDC & Landesschulrat. 2004. *Vorteil Vielfalt: 10 Jahre Minderheitenschulgesetz für das Burgenland – Ein Projekt des Kroatischen Kultur- und Dokumentationszentrum in Zusammenarbeit mit dem Landesschulrat für das Burgenland.*

Holzer, W. ed. 1993. *Trendwende? Sprache und Ethnizität im Burgenland*, Wien: Passagen-Verlag.

Horvath, T. and Müllner, E. eds. 1992. *Hart an der Grenze: Burgenland und Westungarn.* Wien: Verlag für Gesellschaftskritik.

Kolonovits, D. 1996. *Minderheitenschulrecht im Burgenland* (Österreichische Rechtswissenschaftliche Studien, 37), Wien: Manz.

Mayer, S. 2002. *Regionale Europapolitik: Die österreichischen Bundesländer und die europäische Integration – Institutionen, Interessendurchsetzung und Diskurs bis 1998.* Wien: Braumüller.

Nimni, E. ed. 2005. *National cultural autonomy and its contemporary critics*, New York: Routledge.

Offredi, S. 1994/1995. *Le dottrine politiche di Paneuropa: il contributo intellettuale di Richard Coudenhove de Kallergi (1894-1974)*. Universita degli Studi di Milano.

OGM. 1994. *Untersuchung 'Die Kroaten im Burgenland'* (unpublished study, OGM Österreichische Gesellschaft für Marketing for Federal Chancellory and Regional Government Burgenland).

Österreichisches Volksgruppenzentrum. 2000. *Der Report des Österreichischen Volksgruppenzentrums an die drei EU-Weisen* (Webpage of Gesellschaft für bedrohte Völke, http://www.gfbv.it/3dossier/oevz/repoevz.html; Jan 2005).

Pernthaler, P. 2003. "Die Dynamik des österreichischen Minderheitenschutzes". *Europa Ethnica*, 60 (3-4): 75-80.

Pernthaler, P. and Ebensperger, S. 2000. "Der rechtliche Status und die räumliche Verteilung von Minderheiten in den österreichischen Gemeinden im Geltungsbereich der Alpenkonvention". *Europa Ethnica*, 57 (3-4): 117-135.

Rautz, G. 2000. "Die Institution der Volksgruppenbeiräte und mögliche Formen der politischen Vertretung in Österreich". *Europa Ethnica*, 57 (3-4): 136-147.

Rechnungshof 2004. "Volksgruppenförderung", *Wahrnehmungsbericht des Rechnungshofes* 2004/4, p. 3-13.

Riesbeck, P. 1996. *Sozialdemokratie und Minderheitenrecht: der Beitrag der österreichischen Sozialdemokraten Otto Bauer und Karl Renner zum internationalen Minderheitenrecht*. Saarbrücken: Verl. für Entwicklungspolitik Saarbrücken.

Schimmel, E. 2001. *EU-Regionalpolitik im Grenzraum – Intention und Realität: Kooperative Projekte zwischen Österreich und Ungarn*. Technische Universität Wien, Fakultät für Raumplanung und Architektur.

Seger, M. and Beluszky, P. eds. 1993. *Bruchlinie Eiserner Vorhang: Regionalentwicklung im österreichisch-ungarischen Grenzraum (Südburgenland/ Oststeiermark – Westungarn)*. Wien; Köln; Graz: Böhlau.

Stojka, C. 1988. *Wir leben im Verborgenen!*, Wien: Picus-Verlag.

Wagner, P., Kaufmann, A. and Knoflacher, M. 2003. *Halbzeitbewertung des Ziel 1-Programms Burgenland 2000-2006: Endbericht 2003*. Seibersdorf: ARC systems research GmbH.

Zuckerstätter-Semela, R., Hergovich, A. and Puchinger, K. 2001. *Strukturpolitik und Raumplanung in den Regionen an der mitteleuropäischen EU-Außengrenze zur Vorbereitung auf die EU-Osterweiterung – Preparity, Part 13: Die Auswirkungen der EU-Osterweiterung auf Raum- und Zentrenstruktur* (INTERREG IIC, ed. Mayerhofer, P. and Palme, G.). Wien: Österreichisches Institut für Wirtschaftsforschug WIFO.

Ziegerhofer-Prettenthaler, A. 2004. *Botschafter Europas: Richard Nikolaus Coudenhove-Kalergi und die Paneuropa-Bewegung in den zwanziger und dreißiger Jahren*. Wien: Böhlau.

Interviews[70]

(1)   21 March 2005: Press officer of Österreichisches Volksgruppenzentrum.

(2)   22 March 2005: President of Kulturverein österreichischer Roma, district representative of Vienna SPÖ, chairman of Roman Advisory Council.

(3)   23 March 2005: Referee for European affairs, Vienna Business Agency.

(4)   30 March 2005: President of Zentralverband ungarischer Vereine und Organisationen in Österreich & Historian at Austrian Academy of Sciences, member of Hungarian Advisory Council.

(5)   31 March 2005: Chairwoman & Editor of Initiative Minderheiten, coordinators of EU projects.

(6)   7 April 2005: chairman of Österreichisch-Slowakischer Kulturverein.

(7)   8 April 2005: unit for integration and minorities, Wiener Rathaus, MA 17, formerly semi-governmental agency 'Wiener Integrationsfonds'.

(8)   18 April 2005: Historian specialised in Burgenland and minorities; Hungarian nationality.

(9)   20 June 2005: Managing director of Burgenländisch-Kroatisches Zentrum and of Burgenländisch-Kroatischer Kulturverein in Wien, board member of Intitiative Minderheiten, member of Advisory Council for Croatian nationality.

(10) 22 June 2005: Head of Unit for National Minorities, Austrian Federal Chancellory.

(11) 22 June 2005: Municipal Councilor, Chairwoman & speaker on minority questions of Green Party, Vienna Municipal Council.

(12) 22 June 2005, Chief editor of Hrowatski Novine – Kroatische Wochenzeitung, Managing Director of Kroatischer Presseverein.

(13) 24 June 2005, Former leading representative of Carinthian Slovenes.

(14) 27 June 2005, Social and political scientist specialised in minority politics, Austrian Academy of Sciences.

(15) 30 June 2005: written interview response, Parliamentary representative in Landtag Burgenland, FPÖ speaker on minority issues.

(16) 21 July 2005: President of Kroatischer Kulturverein im Burgenland & Member of Croatian Advisory Council.

(17) 21 July 2005: Political secretary of Regional Governor of Burgenland (SP), chairman of Croatian Culture and Documentation Center HKDC, chairman of Croatian Advisory Council and of General Conference of all Advisory Councils.

(18) 21 July 2005: Editor in Croatian Minority Department, ORF Burgenland.

(19) 22 July 2005: parliamentary representative in Burgenland Landtag (Greens), founder of cultural initiative KUGA, Croatian rock-band Bruji.

---

[70]   Anonymous interviews, numerical order according to date of interview.

(20) 25 July 2005: Chairman of Burgenländisch-Ungarischer Kulturverein (recently only section Oberpullendorf), Mayor of Oberpullendorf, Member of Hungarian Advisory Council, former member of Bundesrat (SP).

(21) 25 July 2005: Mayor of border village Bildein, initiator of EU funded cross-border cooperations.

(22) 26 July 2005: Regional Councilor and parlamentary representative in Landtag Burgenland, ÖVP speaker for nationality issues, member of Croatian Advisory Council.

(23) 26 July 2005: Inspector for bilingual schooling in Landesschulrat Burgenland, former chairwoman of Croatian School of Adult Education, member of association ZORA.

(24) 27 July 2005: Managing director of UMIZ – MMIK, Ungarisches Medien und Informationszentrum.

(25) 27 July 2005, Private lawyer & member of Croatian Academics Association.

(26) 28 July 2005: Managing Director of Regionalmanagement Burgenland (development agency managing EU funds).

(27) 28 July 2005: Managing Director of NGO Europahaus Burgenland.

(28) 29 July 2005: Chairman of Verein Roma-Service, member of Roma Advisory Council.

(29) 29 July 2005: President of working group of ÖVP representative of bilingual municipalities in Burgenland, Mayor of Guettenbach, president of Bgld. Gemeindebund, member of Advisory Council for Croatian nationality.

(30) 9 August 2005: secretary of Forum Polonium / Forum der Polen in Österreich.

(31) 9 August 2005: Managing director of association OHO – Offenes Haus Oberwart.

(32) 10 August 2005: Interregionaler Gewerkschaftsrat Burgenland – Westungarn.

(33) 10 August 2005: Former Head of Unit for nationality affairs in Austrian Federal Chancellory, now Federal Ministry of Education, Science and Culture, chairman of Wiener Arbeitsgemeinschaft für Volksgruppenfragen.

(34) 31 August 2005: representative in Federal Parliament, Speaker for Minority issues of Green Party, Burgenland Croatian.

Other Sources

APA: Austrian Press Agency – Defacto Database (electronic news archive).

EFRE and Interreg Statistics, Regionalmanagement Burgenland.

Parliamentary protocols and government reports, www.burgenland.gv.at (Sept 2005).

**Annex 1: Federal subsidies for national minorities
(in 1,000 Euro)**

	1998	*1999*	*2000*	*2001*	*2002*
Croats	1,320	1,158	1,278	1,169	1,196
Slovenes	1,305	1,168	1,123	1,281	1,287
Czechs	344	291	481	478	483
Hungarians	301	276	286	283	278
Roma	239	267	286	399	382
Slowaks	39	31	40	40	40
Other subsidies	287	401	217	5	–
Total	*3,851*	*3,593*	*3,711*	*3,654*	*3,666*

Source: Rechnungshof 2004.

**Annex 2: Location of Burgenland in Austria
and neighboring states**

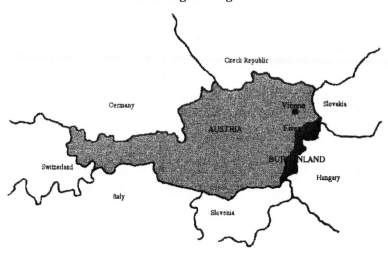

Source: Gmeiner 1999

173

Annex 3: Villages with bilingual population in Burgenland

Sprachenkarte des Burgenlandes - Orte nach der Volkszählung 1991[1]

6 Hornstein/Vorištan*
7 Steinbrunn/Štikapron*
8 Zillingtal/Celindof*
9 Sigleß/Cikleš*
10 Oslip/Uzlop*
11 Trausdorf/Trajštof*
12 Wulkaprodersdorf/* Vulkaprodrštof*
13 Antau/Otava*
14 Siegendorf/Cindrof*
15 Klingenbach/Klimpuh*
16 Zagersdorf/Cogrštof*
17 Draßburg/Rasporak*
18 Baumgarten/Pajngrt*

19 Kaisersdorf/Kalištrof*
20 Weingraben/Bajngrob****
21 Oberpullendorf/Felsőpulya

33 Rauriegel-Allersgraben/ Rorigljin-Širokańi***
34 Mönchmeierhof/Marof***
35 Siget i.d.W./Őrisziget
36 Oberwart/ Felsőőr
37 Unterwart/ Alsőőr
38 Podgoria/ Podgorje***
39 Rumpersdorf/ Rupišče***
40 Allersdorf/ Ključarevci***
41 Spitzzicken/ Hrvatski Ciklin***
42 Podler/Poljanci***

49 Hackerberg/ Stinjački Vrh*
50 Stinatz/Stinjaki*
51 Heugraben/Žarnovica*

Eisenstadt

1 Pama/Bielo Selo*
2 Neudorf b.P./Novo Selo*
3 Parndorf/Pandrof*
4 Frauenkirchen
5 Andau

22 Kleinwarasdorf/Mali Borištof*
23 Großwarasdorf/Veliki Borištof*
24 Nebersdorf/Šuševo*
25 Unterpullendorf/Dolnja Pulja*
26 Nikitsch/Filež*
27 Kroat. Minihof/Mjenovo*
28 Kroat. Geresdorf/Geristof*
29 Großmutschen/Mučindrof*
30 Kleinmutschen/Pervane*
31 Lutzmannsburg
32 Frankenau/Frakanava*

43 Weiden b. Rechnitz/Bandol***
44 Althodis/Stari Hodas***
45 Zuberbach/Sabara**
46 Schachendorf/Čajta**
47 Dürnbach/Vincjet**
48 Schandorf/Čemba**

52 Großpetersdorf
53 Neuberg i.Bgld./Nova Gora*
54 Güttenbach/Pinkovac*
55 Kroat. Tschantschendorf/Hrvatski Čenča*

56 Reinersdorf/Žamar*

57 Heiligenkreuz

Umgangssprache (nach Ortsteilen): ◍ über 50% (VZ 81)

Kroatisch: ○ 2 - 5%   ◑ 5 - 50%   ● über 50%

Ungarisch: □ 2 - 5%   ▨ 5 - 50%   ■ über 50%

* Čakavisch
** Štokavisch
*** Vlahisch
**** Ikavisch- ekavischer Dialekt

Gemeinden mit **Romanes** - Angaben: X

1) Vorläufige Ergebnisse
Quellen: ÖSTAT, VZ 91; Neweklowsky 1978

Source: Holzer & Münz 1991.

PART III

BORDERS, REGIONAL INTEGRATION AND SECURITY

CHAPTER 6

On the Cutting Edge

Border Integration and Security in Europe and North America

Harlan KOFF

I. Introduction

Security issues, such as organized crime, and terrorism have directly raised the visibility of border control on public agendas in Europe and the United States (US). In response to this heightened attention, nation-states have generally followed coercive border control strategies aimed at stopping clandestine migration and human smuggling. One of the most recent measures proposed by the US Congress in December 2005, would entail, should it be implemented, the extension of the current border wall system (in place in four points of the border) to the entire length of the US-Mexico divide. This plan has provoked indignation throughout Latin America where leaders, especially Mexican President Vicente Fox, have decried the measure as "shameful".

Because immigration has been politically framed as a security threat, government responses have increasingly focused on the notion of "control", defined as the prevention of unauthorized migration across state borders. In response, many scholars (i.e. Cornelius) and political actors have questioned the effectiveness of such strategies. In fact, borders have received prominent attention in other political spheres, such as those related to labor market expansion, transnational social movements, environmental conservation, etc. where border cities and regions have historically been recognized as dynamic areas. Many frontiers are known as places of opportunity and expansion, especially when they separate industrialized and developing countries.

This chapter comparatively examines the relationship between immigration and regional integration in Europe and North America through the study of border areas located in both continents. By studying the impact of these global phenomena on sub-national communities, the chapter identifies the mechanisms that best explain the foundations of clandestine migration and its relationship to regional integration models. It does so by responding to three inter-related research questions: 1) how has regional integration affected cross-border cooperation at the sub-national level?; 2) how have recent socio-economic transformations related to cross-border cooperation affected migration regimes in these border communities?; 3) how well have immigration control strategies responded to these recent changes in migration regimes?

Research Design

This chapter focuses on regional integration and the politics of immigration in two border cases, 1) San Diego, USA-Tijuana, Mexico and 2) Bari, Italy-Durres, Albania. These cases are part of a larger study on the impact of regional integration on four border communities. The other two cases in the overall research framework are Northern France-Southern Belgium (area of historic Flanders) and Cucuta, Colombia and San Cristobal, Venezuela. These communities were chosen for this study for specific demographic, political and economic characteristics. In terms of their similarities, all of these cases are traditional migration areas. Second, organized crime has created human smuggling/trafficking rings in each community. Third, each case has historically been considered "integrated" due to binational ethnic ties, economic cooperation or colonial relationships. Finally, regional integration developed in each specific area during the same time period (early 1990s) due to lagging labor markets and the decision by local elites to restructure the foundations of the local economies.

These cases were also chosen due to various traits that make them "most different". San Diego-Tijuana and traditional Flanders represent cases where there is strong asymmetry along the border. In the former, there is little political interaction between the case cities' local governments. Limited collaboration occurs in the economic sphere as chambers of commerce, and planning agencies work to stimulate local markets. In the latter, political cooperation occurs on a daily basis but economic collaboration is lacking due to differences in economic structures. Conversely, demographic and economic symmetry exists between Bari-Durres and Cucuta-San Cristobal. Nonetheless, significant differences remain. Cross-border cooperation between Bari and Durres encompasses both the political and economic arenas as collaboration exists between firms, planning agencies, chambers of commerce, city, provincial and regional governments, and non-governmental organizations. Moreover, because Bari is found in Italy's poorer south, its economy suffers from many of the structural problems found in Durres, such as a large informal economy, high unemployment, limited infrastructure, etc. Cucuta, Colombia and San Cristobal, Venezuela demonstrate little collaboration in either the political or economic arenas due to the deteriorated relationship between the Colombian and Venezuelan governments and the militarization of the border.

Key Concepts

For the purposes of this study, "regional integration" is defined as the cession of sovereignty in policy arenas through multilateral agreements that transfer decision-making powers in those arenas to the supranational level. Specifically, the chapter compares the institutional and legal differences in these arrangements between the European Union (EU) and the North American Free Trade Agreement (NAFTA). "Immigration regimes" are defined not only by the number of immigrants in the case of study, but by the qualitative differences in the types of immigration present in the selected cases. "Clandestine migration" refers to those migrants who cross state borders without proper authorization. "Human smuggling" is defined as the facilitation of unauthorized migration for economic gain, and "Human trafficking" regards the facilitation of unauthorized migration for the purposes of future exploitation in illicit activities.

The direct focus of the study is security in border areas. As in most of the literature on security issues (i.e. Williams), I address this key concept both as "public security" referring to threats to public order and "human security" defined as the maintenance of human integrity or human dignity. Finally, the term "border areas" concerns sub-national communities located in the proximity of nation-state boundaries.

Methods

In order to complete this study, library research on both regional integration (EU and NAFTA) and immigration was carried out at the Robert Schuman Centre for Advanced Studies at the European University Institute, the Center for Comparative Immigration Studies at the University of California, San Diego. Socio-economic data on the sub-national cases was gathered at local universities, local government development agencies, chambers of commerce, and non-governmental organizations (NGOs). Information regarding political cooperation was obtained through personal interviews with representatives of local government and NGOs.

Literature Review

This research project addresses three distinct bodies of literature and attempts to connect inter-related arguments found within them. First, it aims to fill a hole in the emerging literature on border communities.

Numerous scholars, such as Blatter, Maganda, Brooks and Fox, Kirchner, etc. have analyzed the impact of regional integration programs on various aspects of local policy-making in border cities. While these studies are significant, they rarely employ comparative methods. The most extensive book in this field is *Caught in the Middle: Border Communities in the Era of Globalization* edited by Demetrious Papademetriou and Deborah Meyers. This volume presents research on recent transformations in border communities in Europe and North America. Nonetheless, it is a collection of cases and it does not employ the comparative method. Its theoretical introduction, however, accurately presents two arguments: 1) border regions are those most exposed to international phenomena, and 2) local actors often follow policy strategies that contradict national measures. This framework is the basis of this chapter.

The second body of literature that is addressed by this research is that on immigration control. This scholarship has been developed extensively. Authors such as Cornelius, Money, Lavanex, Lahav, and Geddes, to name a few, have analyzed the domestic and international pressures that have led to protectionist policy-making in immigration politics. Nonetheless, one of the weak points of these works is that they often fail to discuss immigration policies within the framework of general political systems. Some authors, such as Holliefield and Bretell, Freeman and Weiner have addressed this deficiency. This chapter aims to contribute to this niche as well.

Finally, the research presented in this chapter addresses issues raised by the current scholarship on comparative regional integration. Authors such as Mattli, Appendini and Bislev, Laursen, and Chambers and Smith have examined regional integration models in comparative perspective discussing issues such as: the national impact of regional integration, the theoretical arguments for regional integration, global incentives for regional integration, and the impact of different institutional frameworks on regional regimes. This chapter aims to address the impact of regional integration on power in sub-national communities in different geographical contexts. By doing so, it addresses an area that is lacking in this international scholarship.

Theoretical Arguments

The chapter is divided into five parts. Following this introduction, part two briefly outlines regional development strategies in Europe and North America and their impact on Mexico and Albania. Part three analyzes the impact of regional integration on the two border case areas. Part four discusses recent border control measures in Europe and North

America and their effects on clandestine migration in the case regions. Finally, part five, presents theoretical conclusions.

The chapter argues that regional integration significantly affects local patterns of cross-border cooperation. However, it also questions the notion that regional integration necessarily creates "border integration". In fact, it contends that levels of border integration vary due to different forms of cross-border cooperation in the economic and political arenas. The chapter then contends that immigration control policies based on the unilateral use of force at the border cannot be effective unless they address the socio-economic contexts, created by regional integration, that significantly affect migration regimes. These socio-economic factors have, in part, been created by cooperation between local governments which have implemented policies aimed at stimulating cross-border economic exchanges. By ignoring this background to migration and focusing simply on border controls, states merely push immigration underground by improving the opportunity structures for organized crime groups that have created networks for human smuggling and human trafficking. The cases presented suggest that developed states need to address the underlying bases of migration if they are to manage this issue with any success. This is best accomplished through multi-level governance and cross-border local cooperation.

II. Regional Integration and Development in Europe and North America

Since World War II, regional integration has taken many forms. These models have been thoroughly discussed in the comparative literature on regionalism (see Laursen, 2003, Mattli, 1999, Appendini and Bislev, 1999). Rather than focus on the theoretical debates that surround regional integration, the purpose of this section is to briefly present the evolving contexts in which border politics have developed in Europe and North America. By describing these models of integration and their corresponding development strategies in Mexico and Albania, this section provides the necessary background for the analysis presented in part three.

The European Union and the North American Free Trade Agreement Compared

Stephen Clarkson (2000) asserts that comparing the European Union (EU) and the North American Free Trade Agreement (NAFTA) is like comparing apples and oranges.[1] When the Treaty of Rome was signed in 1957, European integration was based on the tremendous need to reconstruct the continent, as well as political idealism tied to notions of international peace and cooperation. For this reason, the concept benefited from a tacit consensus throughout Europe allowing the forefathers of European integration to lay the foundations for integration in the political, economic and social spheres.

Since that time, integration objectives have shifted away from responses to security issues and the implementation of ideals concerning peace, and moved towards the efficient management of economic forces in the global economy. As regional integration developed in Europe and other parts of the world, it gradually became linked more closely with political and economic competitiveness.

This is visible in the differences between the EU and NAFTA. Because the EU represents a model of early integration, its institutional and legal mechanisms are more developed and its political objectives are more ambitious than those included in NAFTA, an example of later, market-based integration. These differences are summarized in Table 1.

Table 1. EU and NAFTA Compared

	Characteristics of Regional Integration			
	State-Market Relationship	*Development of Regional Institutions*	*Common Identity*	*Commitment to Harmonization*
EU	Balanced	Economic/Political	Medium-High	High
NAFTA	Neoliberal	Economic-Regulatory	Low	Low

European and North American Development Strategies

The political and economic disparities between the EU and NAFTA are clearly evident in the development strategies that they promote. The greatest difference between the economic aspects of regional integration in Europe and North America regard the notion of equity. In North America, regional harmonization is not an objective of NAFTA and socio-economic equity has never been part of the agreement. Conversely, European integration has institutionalized the principle of equity though regional funding and social programs aimed at reducing "the differences between the various regions and the backwardness of

---

[1]   Clarkson, S. 2000. *Apples and Oranges: Prospects for the Comparative Analysis of the EU and NAFTA as Continental System.* Robert Schuman Centre Working Paper 2000/23. Florence: European University Institute.

the less favoured regions.[2]" EU strategies focus on reducing disparities in the economic, political and social spheres, while NAFTA virtually ignores the latter two arenas (see Table 2).

Table 2. Commitment to Development

	Economic Markets	Political Institutions	Civil Society
EU	High	High	High
NAFTA	High	Medium (Peripheral Interest)	Low

The impact of these differences has been demonstrated by many scholars of development, such as Pastor, Hansen, and Woodruff. Integration in North America has weakened the relative positions of peripheral areas whereas integration in Europe has strengthened them. A comparison between the strongest and weakest national economies in each regional system shows that the Gross Domestic Product (GDP) of the United States is 6.5 times higher than that of Mexico whereas the GDP of Germany is only 2.4 times that of Portugal. Moreover, within Mexico, per capita income within the poorer southern states is 62% less than that found in the wealthier northeast.[3] The impacts of these differing approaches to development are evident in recent economic trends in Mexico and Albania.

NAFTA and Mexico and the EU and Albania

Scholars of development, such as Larry Diamond, have long identified three arenas that contribute to stability and prosperity: market liberalization, the presence of transparent political institutions and fair legal structures, and cohesive civil society. The greatest difference between regional integration in Europe and North America regards the emphasis placed on the various sectors of development.

Under the auspices of NAFTA, it is clearly evident that political reform is not a priority. Instead, it is a secondary objective of the agreement to be achieved through economic reform. This outcome occurred for two reasons. First, as stated earlier, NAFTA was championed by US corporations with little interest in political reform. Due to these external restraints, US leaders (especially President George Herbert Bush), had little opportunity to include democratization on the NAFTA agenda. Second, Mexican leaders also opposed a dual agenda. When the *Partido*

---

[2]   Pastor, R. A. 2002. "A Regional Development Policy for North America: Adapting the European Union Model", in Edward, Chambers and Smith. eds. *NAFTA in the New Millenium*. La Jolla: Center for US-Mexican Studies, p. 397-424, at p. 398.

[3]   *Ibid.* p. 401-02.

Revolucionario Institucional (PRI) began liberalizing the economy after the shocks of the early and mid 1980s, democratic reform did not follow. Instead, PRI leaders maintained oligarchic political structures that permitted them to remain in power. Political reform did not even become part of Mexico's agenda until Vicente Fox and the *Partido del Accion Nacional* (PAN) won the 2000 presidential election, a full eight years after NAFTA was signed.

The results of this approach have been clear in Mexico. Due to trade liberalization, many economic indicators demonstrate improvements in Mexico's overall economy. Foreign investment has increased (stock of foreign investment doubled between 1994 and 1999 from US$42 billion to US$83 billion), as has gross domestic product (to one trillion US dollars), exports in manufacturing (from US$41.6 billion to US$105.9 billion), and employment (official unemployment rate decreased from 3.7% in 1993 to 2.3% in 1999). However, social marginalization has also boomed as precarious labor has led to a relative decline in wages (by 13.2%) and regional disparities in economic performance have also grown (25 of 32 states account for less than 3% each of total production in Mexico).[4] Politically, reform of the justice system and the institutionalization of human rights have been slow to develop. Furthermore corruption continues to run rampant even after the change in power in 2000. Public dissatisfaction with President Fox combined with some high profile scandals that have included ruling PAN officials have actually made a quick return to power for the PRI a legitimate possibility in the next presidential elections. This is especially true given the PAN and PRI's successful campaign to destitute the leading Left-wing candidate for President, Andres Manuel Lopez Obrador from his position as Mayor of Mexico City. This act has created mass public outcry and instigated international criticism that it is a blow to democracy in Mexico.

Conversely, despite numerous challenges to development, democracy in Albania has been consolidated more evenly and international organizations, such as the World Bank, Council of Europe and the EU have praised the country for the progress that has been accomplished both politically and economically. For example, Nadir Mohammed, World Bank Country Manager for Albania has stated:

> We are encouraged by this performance and commend the government of Albania for its commitment. Major regulatory reforms have put the banking sector on a firmer footing, energy-pricing reforms are reducing the fiscal

---

4   World Bank Group. 2005. Mexico Data Profile.

burden, and privatisation is picking up momentum. Albania is also making important progress in its path towards EU integration.[5]

In fact, since the fall of the Communist regime in 1991, Albania's economic markets and political institutions have progressively stabilized, despite significant internal and external challenges, such as the 1997 pension crisis and the military conflicts in Kosovo and Macedonia. While these events created added difficulties for the development process, they also led to heightened foreign aid and political support. Unlike Mexico, Albania has benefited from significant foreign assistance directed at political and social incorporation, especially from the EU. Since 1991, Albania has received over one billion euros in EU assistance.[6] At the beginning of the 1990s, financial aid targeted food and emergency relief. Since then, the EU has heavily subsidized economic and political reform through the PHARE (*Pologne et Hongrie: Assistance pour la Restructuration Economique*) and CARDS (Community Assistance for Reconstruction, Development and Stabilization) programs aimed at: developing a fair judicial system and transparent legal structures, establishing basic infrastructure, improving customs and tax collection, developing civil society, strengthening education and environmental protection, and opening Albania to its neighbours through transportation programs and cultural and educational exchanges.

These investments have significantly improved Albania's political, economic and social situations. Following the fall of Communism, the country was confronted with deep social divisions, institutional inefficiency and underdeveloped markets. Gains in all three arenas have been evident. Politically, a stable parliamentary republic has been established, public administration has been improved dramatically, and a fair judicial system has begun to develop (which remains one of Mexico's greatest problems). One of Albania's largest institutional challenges has been the development of an effective tax and customs administration. Because of the progress accomplished in this area, tax revenues increased 26% in 2000 with respect to 1999 which contributed to a decrease in the fiscal deficit from 11.5% of GDP to 9.2%.[7]

Other economic indicators reflect the positive effects of political development on economic markets. Overall, Albania has shown steady growth since 1991 (except for 1997 due to the aforementioned pension

---

[5] World Bank Group. 2005. World Bank Assesses Albania's Economy, Outlines Reform Challenge.

[6] European Commission Delegation to the Republic of Albania, Assistance. 2005. www.delalb.cec.eu.int/ep/eu [accessed June 2005].

[7] European Commission: External Relations Directorate General. 2002. Albania: Country Strategy Paper: 2002-2006, p. 11.

scandal) with yearly improvements of seven to 8% of the country's overall GDP.[8] The service sector has the most dynamic growth rate at 20.1%, followed by construction (18.4%) and industry (13.6%). Foreign direct investment has also increased from US$20 million in 1992 to US$135 million in 2002 and inflation is at its lowest point since the economic transition began fourteen years ago.[9] Finally, remittances from Albanian's living abroad are estimated at approximately US$440 million.[10]

This final statistic, however, touches on many of the social problems that continue to afflict Albania. Due to the weakness of civil society, the importance of poverty, the structural problems in education, the strength of organized crime, and low wages, many Albanians have chosen to emigrate. It is estimated that 500,000-600,000 Albanians left the country in the 1990s, creating a problematic "brain drain". Moreover, approximately 36,000 migrants pass through the country each year on their way to the European Union.[11] Albania has become of the busiest transit points in Europe for human smugglers and traffickers. For this reason, special attention has been placed on police training and border controls. In fact, after having been considered one of the "most closed" countries in Europe throughout the Communist regime of Enver Hoxha, Albania has become one of the continent's most open, due to its geographic position. For this reason, border issues have become as publicly prominent as they have traditionally been in Mexico. This is the focus of the section three.

III. The Impact of Regional Integration on Border Communities

Since the early 1990s, asymmetry and shifts in balances/uses of power have been a central focus of border studies. Many scholars (i.e. Blatter, 2004, Papademetriou, 2001, Kirchner, 1998) have noted that local actors often follow economic and political strategies that counteract policies instituted at the national level. Moreover, excluded groups on both sides of a border often unify their efforts to further a common political cause (i.e. Brooks and Fox, 2002). What has not been sufficiently examined is what types of cross-border cooperation develop at the local level from regional integration and how these varying models of political cooperation relate to the management and restructuring of

---

8 World Bank Group. 2005. Albania Data Profile.

9 Albanian Economic Development Agency, http://aeda.gov.al/achievements.htm, [accessed June 2005].

10 European Commission. 2002. Albania: Country Strategy Paper: 2002-2006, p. 12.

11 *Ibid*, p. 10.

economic markets. These questions have not been analyzed closely in much of the literature. For example, Joachim Blatter correctly argues that scholars of borders often assume two theoretical points related to globalization: 1) that political economy represents the increasing integration of the political and economic spheres and 2) regional integration creates cross-border cooperation by definition.

This section examines border politics and markets in San Diego-Tijuana and Bari-Durres. Rather than simply focusing on how integration has occurred in these communities, this section questions whether integration has occurred at all, and if so, to what extent? Its basic premise is that border integration involves a series of changing relationships that encompass markets and political systems. These relationships may shorten or lengthen the distances between communities, but these communities remain separate entities within each case of study. By focusing on the character of the relationships that have developed between local political institutions, as well as the nature of cross-border economic ties between cities, this section attempts to describe how the regional integration models described above have influenced these border communities. These outcomes will then be utilized as the basis of the analysis of immigration policies presented in part four.

Defining Border Integration: A Qualitative Question

Globalization has certainly increased the number of cross-border transactions and it has accelerated rates of economic exchange. Urban planners interested in border questions often examine this latter aspect of globalization. For example, Steven P. Erie, one of the leading scholars of the San Diego-Tijuana region's economic transformation, has publicly called for improved transportation infrastructure on both sides of the border because current capabilities are insufficient for faster exchanges of commodities and services. Similarly, scholars of economic development in the Adriatic Basin (which includes Bari and Durres), such as Gianfranco Viesti, have acknowledged that increased economic interdependence in this geographic area has created a need for improved airports, seaports, and roads.

While globalization has certainly affected the volume and speed of economic exchanges, it cannot be taken for granted that it has created market integration in border areas. Integration, defined as the incorporation of market structures into (even loosely) unified economic bodies, is not found in all border communities. Through an analysis of socio-economic changes in San Diego-Tijuana and Bari-Durres, this section aims to show that communities in border areas do not necessarily need to be considered integrated, even if their patterns of development are similar. Moreover, these cases show that regional integration models

187

have an impact on how political and economic cooperation occurs at the local level.

San Diego-Tijuana

The San Diego-Tijuana area represents the busiest border in the world with about 70 million crossings each year. The region's economy is worth US$100 billion which ranks it as the thirty sixth largest economy in the world. Its population has skyrocketed to over 4 million inhabitants, 2.8 million of whom reside in San Diego and 1.2 million of whom live in Tijuana.[12] For these reasons, this case is often considered a "well-integrated" border. In fact, this demographic and economic growth has been significantly affected by NAFTA as the importance of this area developed with the introduction of regional integration. Following World War II, both San Diego and Tijuana were only minor cities with limited industrial bases. Due to the region's climate and location, both cities received significant amounts of internal migrants. Moreover, San Diego's economy received a boost from the defense industry, as Navy and Marine bases located in the county were expanded. Nonetheless, the signing of NAFTA in 1993 accelerated expansion tremendously.

Since that year, both of these cities experienced major transformations of their economic foundations. Between 1990 and 1992, 53,000 jobs were lost in San Diego County.[13] This represented 5% of the local workforce. Because of this recession, local officials effectively planned and negotiated a shift to a "New Economy" based on high technology. They identified nine clusters in high-tech fields, including biomedical production, biotechnology and pharmaceuticals, computer software, wireless telecommunications, electronic and computer manufacturing, business services, financial services, defense and transportation manufacturing, and environmental technology. This shift fundamentally changed the focus of the San Diego economy from one based on regional markets to an economy oriented towards international exports. Not surprisingly, due to the timing of this transformation as well as San Diego's geographic location, Mexico became the primary target of these exports. Whereas other California cities, notably Los Angeles and San Francisco have fostered east-west connections between Asia and

---

[12]   Blatter, J. 2004. "From Spaces of Place' to 'Spaces of Flows'? Territorial and Functional Governance in Cross-Border Regions in Europe and North America". *International Journal of Urban and Regional Research*. 28, 3 (Sept.), p. 536.

[13]   Alarcon, R. 2005. "Mexican Migration Flows in Tijuana-San Diego in a Context of Economic Uncertainty", in Kiy, R. and Woodruff, C. eds. *The Ties that Bind Us: Mexican Immigrants in San Diego County*. La Jolla: Center for US-Mexican Studies, UCSD.

Europe, San Diego's ties run north-south. This is obvious in its trade with Mexico which receives 44% of the city's exports, compared to 23% for all Asian states, and 9% for Canada. Furthermore, of the 126,000 international passengers who pass through San Diego's airport every year, 102,000 are estimated to be Mexican.[14]

Similarly, Tijuana has greatly benefited from NAFTA. The city has an official growth rate of 6.9% each year. Its population has increased from 740,000 in the late 1970s to 1.3 million in 2000. Actually, local planners expect growth to continue at even higher rates doubling the population again by 2020.[15] This demographic expansion has significantly contributed to the development of the local *maquiladora* economy.[16] Even though there is much internal migration to Tijuana from poorer parts of Mexico, the city has a highly educated workforce. The official literacy rate is 97%. The local Chamber of Commerce has utilized high rates of education to lure companies such as SONY, Panasonic, and Samsung to the area. In fact, Tijuana has one of the highest concentrations of specialized workers in Mexico and the highest concentration of *maquiladoras* in the country.[17]

Of course, the city's economic growth has also been aided by its geographic position. Tijuana has proclaimed itself "the television capital of the world" because of the strong presence of *maquiladoras*. These televisions, radios, stereos, etc. are mainly produced for the US market. They represent the leading imports from Mexico that cross into the US at the San Diego border. Moreover, in 2002, they were 48.6% of the all exports from Mexico with a value of US$18,680 million.[18]

While these trends that have enveloped San Diego and Tijuana are encouraging, they cannot be considered the fruit of cross-border cooperation. In fact, they closely reflect the market-based model, described above, that was implemented under NAFTA. Little cross-border governmental cooperation between these cities exists. A planning agency called SANDAG (San Diego Association of Governments) has instituted a Border Committee which formally includes officials from all of the border cities in Southern California (including San Diego) and

---

[14] Erie, S. P. 1999. "Toward a Trade Infrastructure Strategy for the San Diego/Tijuana Region". Briefing Paper, San Diego Association of Governments, Feb.

[15] Tijuana Economic Development Corporation. 2005. City of Tijuana. http://www.tijuana-edc.com/deitac/ [accessed June 2005].

[16] *Maquiladoras* are bonded assembly plants found in Mexican border cities that are permitted to import goods without payment of import duties. These goods, especially electronics, are further processed or manufactured and exported. When the goods enter the US, tariff is levied only on the value added outside the US.

[17] Tijuana Economic Development Corporation.

[18] Erie. "Toward a Trade Infrastructure Strategy for the San Diego/Tijuana Region".

Northern Baja California (including Tijuana). However, a member of this committee stated during a personal interview that the Mexican representatives do not actively participate and their input is largely symbolic. Most scholars of this border, (i.e. Carmen Maganda, 2005) point to two obstacles that have blocked cross-border cooperation. First, the asymmetry of power at the border offers little incentive for leaders in San Diego to enter into institutionalized dialogue with their Mexican counterparts. Economically, San Diego is ten times wealthier than Tijuana (added value of US$90 billion *versus* US$11 billion[19]). For this reason, there have actually been anti-border reactions amongst the electorate in San Diego when integration has been discussed. Because local leaders lose more votes that they gain by opening new avenues of cooperation, they avoid such dialogue.

Leaders in Tijuana have expressed equal concern about sovereignty in negotiations with counterparts from San Diego. Due to the exploitative behavior of US officials in the past, many Mexican leaders, especially those in border communities, worry about ceding sovereignty in political and economic matters. Furthermore, because there is significant anti-US sentiment in the area, many leaders worry about the impact of cross-border collaboration on their own re-election chances. For its part, the Mexican Consulate in San Diego has focused its energies on protecting the rights of Mexican migrants in the city who, due to aggressive behavior by local immigration officials, often risk deportation when their situations are not regular. For this reason it has had limited involvement in cross-border cooperation.

This situation demonstrates numerous difficulties that have prevented political collaboration between San Diego and Tijuana. In fact, due to the lack of cooperation in the political sphere, the above-mentioned economic sectors have expanded without integration into a cross-border market. Instead of a single, integrated economy one finds two separate parallel markets. Tijuana does export goods to the United States, but San Diego is not considered a consumer market. It is merely a point of passage to other areas in California and beyond, especially Los Angeles, with whom Tijuana has tighter political and trade relationships. Similarly, San Diego's high-tech products are destined for markets in Mexico's interior, by-passing Tijuana. These facts suggest that consumption is a much better indicator of integration than production.

---

[19] Anguiano-Tellez, M. E. 2005. "Cross-Border Interactions: Population and Labor Market in Tijuana", in Kiy, R. and Woodruff, C. eds. *The Ties that Bind Us: Mexican Immigrants in San Diego County*. La Jolla: Center for US-Mexican Studies, UCSD.

A second observation from the San Diego-Tijuana case concerning the impact of supranational integration regards changes internal to each city's socio-economic structures. Because market expansion occurred on both sides of the border without significant political collaboration beyond deregulation, the populations of each of these border cities have been exposed to economic fluctuations caused by regional integration. Due to the lack of concern for equity, increased social stratification has occurred in both cities. For example, despite overall increases in levels of wealth in San Diego, social marginalization has risen as well. Rafael Alarcon (2005) reports that from 1990 to 1998, job growth in the non-targeted industry clusters (71,079 positions) far outdistanced that in the targeted industry clusters (36,794).[20] Similarly, the Union Tribune (2004), the city's newspaper, reported that even though unemployment had decreased below 6%, most of the positions that encouraged this drop were in precarious or part-time employment which has led to a resulting decrease in wages and quality of life, especially among ethnic minorities, including Mexicans and Mexican-Americans. In fact, geographical studies of the city's neighborhoods (i.e. San Diego Association of Governments, 2004) show a distinct separation between wealthy districts and poorer ones. Census data gathered at the neighborhood level shows that this gap has significantly increased between 1990 and 2000.

Tijuana has faced a similar and even more dramatic situation. Despite an increase in overall levels of wealth, education and health, wages have been decreasing in the area. In 1993, when the *maquiladoras* first opened in the city, the average wage was US$1.80/hour. This wage has dropped to US$1.70/hour since then. Moreover, due to less expensive international competition, Tijuana is losing jobs to other parts of the world, most notably China, where wages are almost four times cheaper (less than fifty US cents per hour). Between 2000 and 2002, the city's *maquiladora* industry cut over 300,000 jobs.[21] This loss of employment; combined with lower wages, and dynamic demographic growth has been a dangerous combination for Tijuana. Entire abusive neighborhoods, constructed by residents without government authorization and usually lacking basic hygienic services, have been erected by internal migrants from Mexico's poorer southern states. City planners estimate a need for 50,000 new low-cost homes to be built within the year with the same necessity next year.[22] This housing shortage has also led to in-

---

[20] Alarcón. "Mexican Migration Flows in Tijuana-San Diego in a Context of Economic Uncertainty".

[21] Tijuana Economic Development Corporation.

[22] *San Diego Union Tribune*, 2005. www.SignOnSanDiego.com [accessed 6 January 2006]

creased crime and violence. Thus, economic and social trends in Tijuana mirror those found in San Diego. This can due be attributed to market forces promoted by NAFTA, which have created marginalization in both border communities despite the lack of formal border integration, such as that which has developed in Bari-Durres.

Bari-Durres

Unlike, the previous North American case, one finds numerous examples of political and economic collaboration between Bari, Italy and Durres, Albania. In fact, the City of Bari has proclaimed itself "A Bridge Between East and West". Located in the Southern Italian region of Puglia along the Adriatic Sea, the city's economic and political ties have, in fact, been traditionally stronger with counterparts in the Balkans than they have with Northern Italian cities. Since the fall of Communism in Albania and the ex-Yugoslavia, city, provincial and regional officials have focused on re-establishing these historical links in order to improve the area's economic foundations, which were shaking in the late 1980's and early 1990s. One of the closest relationships that has been resuscitated is that with the Albanian city of Durres. Like Bari, it is an Adriatic port city with a rich history. Due to Albania's colonial situations, it has passed from one ruler to another thirty-three times. During the early part of the century, it was a focal point of the Albanian independence movement, in part due to the importance of Italian influences.

The economy of Durres has been built around trade in the Adriatic Basin and the city's recent history has been marked by its relationship with Italy. In fact, Italians have had a strong interest in the city since the end of the 19th century. For this reason, they significantly invested in the independence movement while Albania was ruled by the Austro-Hungarians. During the brief rule of King Zog in the 1920s, Italy, and specifically Puglia, contributed large amounts of capital to improve infrastructure in the city. Specifically, investments were made to construct wider roads, improve the port, and build factories that produced predominantly flour, cigarettes and pasta. Because of these interests, the city has been occupied twice by the Italian military, first in 1918, and later under Mussolini's Fascist regime in the 1930s.

Like San Diego and Tijuana, Bari and Durres underwent significant socio-economic changes during a period of regional integration due to the consolidation of the European Union in 1992. The fall of Communism in Albania (1991-1992) coincided with the *tangentopoli* corruption scandals that fundamentally altered Italian politics and significantly affected Bari. While Durres was learning to cope with a market economy, Bari was addressing the loss of jobs at its three biggest employers:

the *Aquedotto Pugliese*, the seaport, and the *Fiera del Levante*, which were semi-public entities that had been mismanaged and defrauded through corruption.

This situation created economic need on both sides of this water border. Durres, one of the most dynamic cities in Europe's poorest country, and Bari, a poorer Italian city with great potential but structural problems, constructed a partnership based on reciprocal interests related to regional integration. In fact, these cities have created a working relationship through the private sector, as well as through political collaboration on European Union-sponsored programs. This has led to a continuous working relationship in various sectors that represents a model of stable political and economic border integration.

In fact, there is great symmetry in the relationship between these two cities. Demographically, Bari is the larger metropolitan area with approximately 500,000 inhabitants compared to 350,000 people living in the Province of Durres and 180,000 in the city, itself. Both cities rely heavily on foreign trade through their ports. More than 2 million tons of merchandise passes between them each year, which represents eighty 5% of Albania's total import-export.[23]

Both cities have a diversified economy based on the dominance of small firms in various economic sectors. Of the 16,501 businesses present in Bari 89.1% employ five people or less.[24] Similarly, 78.5% of the 5,844 companies active in Durres employ less than ten people.[25] Moreover, in both of these cities, these businesses are spread through different economic sectors. In Bari, 45% are found in industry, 33.5% in commerce and 21.5% in services.[26] Outside the city, there is a strong agricultural sector based on seasonal labor and small farms. In Durres, 8.9% of the companies are industrial, 49.9% are commercial, and 35.8% are active in services or transportation.[27]

A second structural similarity between Bari and Durres is their reliance on "traditional family labor". Under this model, the family units that live in the areas surrounding these cities are not dependent on a single breadwinner. They rely on numerous components of the family working part-time or precarious jobs. This makes employment flexibility a necessity in order to guarantee the security of entire families. Potential for mobility within the local workforce is high in both sectoral

---

23 Official Statistics. City Government of Durres. Website. http://www.tavolovirtuale. tno.it/durazzo.asp [accessed June 2005].

24 ISTAT. 1996. *Puglia: Informazioni Utili.* Rome: ISTAT.

25 Official Statistics. City Government of Durres.

26 ISTAT. 1996. *Puglia: Informazioni Utili.* Rome: ISTAT.

27 Official Statistics. City Government of Durres.

and geographic terms. Giandomenico Amendola writes: "Its [the family's] flexibility has acted as a buffer during periods of economic crisis... in that this labor, which outperforms the urban and less reusable workforce, was easily displaced, without a heavy impact on either the local system of production or, above all, the political system[28]". The negative impact of this labor model, however, has been the lack of incentives to invest in technology or improved production techniques.

For these reasons, Bari's economy has never realized its full potential. Unemployment rates remain high at 16.7%. This figure is especially dramatic for women (26%) and young people (33.7%). Moreover, education levels are low as only 19.3% of the city's working population has a high school or university degree.[29] This has contributed to an increase in precarious work and the development of a significant informal economy that includes activities that are "not regulated by the institutions of society in a legal and social area where similar activities are regulated[30]".

Because the size of informal markets is a function of flexibility in economic systems, they are prominent in both Bari and Durres. Since 1999, more than 1,000 businesses in Bari are denounced for "irregularities" every year.[31] The informal sector is so important that one local businessman who operates in a small town in the suburbs of Bari reports: "I am regular and I pay my workers' contributions and my taxes. However, I cannot report all of my sales because I would lose two-thirds of my clients. Unfortunately, if I do not sell to these firms, I cannot make a living myself. This is the market reality[32]". A similar situation exists in Durres as Albania's regulatory agencies are not stable.

Because of their history of cooperation and present economic symmetry, Bari and Durres have become partners in the regional integration of the Adriatic Basin. Numerous protocols have been signed by city, provincial and regional officials creating economic partnerships and cultural/educational exchanges. The aforementioned *Fiera del Levante* has organized business fairs in Albania aimed at promoting cross-border trade and the *Gazetta del Mezzogiorno*, Bari's main newspaper has begun publishing special editions in Albanian.

---

[28] Amendola, G. 1985. "Segni e Evidenze", in Amendola G. *et al.* eds. *Segni & Evidenze*. Bari: edizioni Dedalo, p. 9-23, at p.12.

[29] ISTAT. 1996. *Puglia: Informazioni Utili.* Rome: ISTAT.

[30] Castells, M. and Portes, A. 1989. "World Underneath: the Origins, Dynamics, Effects of the Informal Economy", in Portes, A. ed., *The Informal Economy*. Baltimore: The Johns Hopkins University Press, p. 12.

[31] Official Statistics. 2000. Ispettorato Provinciale del Lavoro.

[32] Personal interview with author conducted 10 October 2001.

The largest benefit of this partnership has been the arrival of European Union funding. Both Bari and Durres have received EU structural funds to improve their airports, modernize their seaports and build roads. The largest sum of thirty-three million Euros was contributed within the framework of the INTERREG program. The cities were also chosen for participation in the Corridor 8 program aimed at constructing rail and auto routes "from Northern Europe to the Black Sea". This large endeavor will eventually connect Bari and Durres with the cities of Tirana, Skopje, Sofia, Plovdiv, Burgas and Varna. Thus, Bari and Durres have the opportunity to become transportation nodes between Western and Southeastern Europe. This demonstrates a true interest in border integration, politically, economically and socially. In fact, EU support has not only affected infrastructure and transportation. While overall unemployment rates remain high compared to San Diego-Tijuana, they have decreased over the last ten years. This reduction of social marginalization has significantly affected migration in the area. This point is further developed in the discussions below.

IV. Contemporary Developments in Border Control Strategies

Like the literature on regional integration, immigration studies pay significant attention to borders, but they rarely focus comparatively on border communities themselves. Authors, such as Alarcon, Kopinak, Andreas, and Alegria, have extensively studied local immigration politics along the US-Mexico border. Similarly, locally focused works (by scholars such as Di Comite, Pace, and Bonifazi) have discussed border immigration politics in Europe. Despite these similarities, little dialogue exists between these bodies of literature. This chapter aims to create this link. Local border immigration studies on both European and North American cases accurately argue that while migration discussions affect national political debates, border communities are the areas where international developments and local citizen concerns intersect. In the sub-field of clandestine migration, most attention is paid to discussions pitting public security (migration controls) *versus* human security (rights of migrants) in these areas.

By comparatively focusing on recent developments in border areas, this chapter aims to add a new dimension to discussions of migration politics. On one hand, the evidence presented in the previous section suggests that local and regional actors follow border strategies that directly contradict control policies orchestrated at the national level. Instead, border integration follows models of regional integration. Nonetheless, the effects of the logic of control which lies at the base of national migration policies in Europe and North America are clearly visible in the border cases discussed in this chapter.

Controlling Borders in the EU and North America

The relationship between development and immigration has already been illustrated in the literature on international migration (see Cornelius *et al.* 2004; Jamieson and Silj, 1998). Nonetheless, immigration policies in Europe and North America largely ignore the relationship between these phenomena. In the European Union, despite the activities of the European Parliament and the Commission in areas related to social integration and anti-discrimination, the Council, which has ultimate policy-making authority in this arena, has focused its attention almost entirely on border controls.

Since the 2002 Seville Council meeting, the EU has further strengthened its efforts to control its external borders. The resolutions passed at this meeting further reinforced the EU's commitment to improving border security. The approved strategies entail the following:

A) *Harmonizing measures to combat illegal migration*: including the creation of a common visa identification system; acceleration of the conclusion of readmission agreements with specific countries identified by the Council; approval for elements of a program on expulsion and repatriation policies, including the optimization of accelerated repatriations to Afghanistan; and formal approval for reinforcing the framework for suppression of assistance for illegal migration.

B) *Progressive operationalization of coordinated and integrated administration of external borders*: including joint operations at external borders and the creation of government liaison officials for immigration; drafting a common model of risk analysis; drafting common training procedures for border police together with consolidation of European norms concerning borders; and drafting a study by the Commission on the administration of external borders.

C) *Integration of immigration policy in the relations of the Union with third countries*: including a provision that states that "a clause be included concerning the common administration of migratory flows and regarding obligatory readmission in the case of illegal immigration in all future agreements of cooperation, association or the equivalent that the European Union or the European Community signs with any country[33]"; and a systematic evaluation of relations with third countries that do not collaborate in the fight against illegal immigration.

---

[33] Teresa Guardia (Ministry of the Interior Spain), 2002. "New Strategies In the Fight Against Trafficking in Human Beings During the Spanish Presidency of 2002". Madrid: Sept.

Most recently, many EU states have supported the creation of transit camps in third states, such as Libya, so that clandestine migrants can be intercepted before they reach Europe's borders, and in June 2005, the EU harmonized forced deportation procedures.

The results of this strategy are clearly evident in the Adriatic Basin. Along the shores of the Region of Puglia, the Italian Coast Guard carries out 28,000 hours of navigation and 1,500 helicopter missions each year. It also boards approximately 3,500 vessels yearly searching for clandestine migrants.[34]

Immigration policies in the United States have followed a similar pattern. Despite provisions for internal migration controls (i.e. employer sanctions) that were included in the 1996 Immigration Reform and Control Act, most of the policy changes in the field of immigration have focused on preventing the infiltration of the country's external borders. This is especially true at the border with Mexico where the "Southwest strategy" instituted by President Clinton in 1993 and strengthened by President George Walker Bush, has created concentrated efforts to regulate border crossings and stop clandestine migration. Under this strategy, the budget for the Immigration and Naturalization Service more than tripled, and the Border Patrol has more than doubled in size, making it the second largest federal law enforcement agency. Most significantly, this strategy has included the construction of three 66 mile walls at strategic points of the border, including San Diego-Tijuana, forcing clandestine migrants to cross in areas where geographic conditions are harsh including deserts, and extremely cold mountainous terrain.

Effects of Border Controls

Since the implementation of more rigid border controls in North America and Europe, illegal migration has obviously not stopped. The current stock of clandestine migrants in Europe is estimated to be over one million. In the Unites States this stock is approximately 9.3 million, which is 26% of the foreign-born population, and 5% of the total workforce. It is estimated that 300,000 to 400,000 undocumented migrants enter the US every year.[35]

While these strategies have not stopped migration, they have changed its character significantly. Scholars of migration both in the Mediterranean and in North America have noted that these measures have severely restricted short-term commuter migration and cyclical

---

[34] Personal Interview with Italian Coast Guard Official conducted 10 October 2001.

[35] Cornelius, W. *et al.* eds. 2004. *Controlling Immigration*. Stanford, CA: Stanford University Press, p. 59-60.

seasonal migration. Whereas legal short-term migration (often daily) was once common in border areas, it now represents just a small amount of what it used to be. Moreover, anthropologists conducting work with clandestine migrants (i.e. Chavez, 1998) have shown that these workers have reduced their number of yearly border crossings for fear of being discovered upon re-entry. This has curtailed return migration as well.

The biggest impact of these policy developments has been the increase of danger in the migration experience. Paradoxically, border controls have significantly reduced the safest and economically most productive migration flows while they have stimulated activities in more dangerous clandestine migration regimes. Because immigration controls have become sophisticated, human smugglers have been forced to further professionalize their networks, increasing transaction costs. For this reason, the cost for passage into Europe or the United States has tripled in the last decade. Moreover, because smugglers have invested so much in these networks, they often combine this business with other criminal activities, such as drug trafficking and the transport of arms or stolen vehicles. Finally, the de-humanization of smuggling through its growth into a multi-national business has led to increased amounts of human trafficking for forced labor or the sex industry.

These assertions are supported by statistics gathered in the border communities included in this study, which illustrate the increasingly dangerous nature of contemporary migration experiences. Between 1993 and 2003, more than four hundred people have died attempting to cross into California from Baja California.[36] Similarly, almost three hundred people died during their crossing from Albania into Puglia during the 1990s.[37] Thus, increased attention to border controls have contributed to systemic transformations in migration regimes that have created increased threats to both public and human security, suggesting the need for an alternative approach to the management of migration flows. Recent developments in border communities suggest that multi-level, binational cooperation could provide an effective strategy to the management of migration. This is the focus of part five, the conclusion.

V. Conclusion: Improving Border Control Through Border Integration

In both political and academic debates concerning regional integration and immigration, it is generally assumed that these phenomena weaken the ability of nation-states to control their external borders.

---

[36] Operation Gatekeeper Factsheet. 2002. www.crlaf.org/factsht.htm.

[37] Personal Interview with Italian Coast Guard Official conducted 10 October 2001.

Recent developments in both European and North American politics demonstrate the fear that this assumption can create amongst native citizens. For example, in June 2005 a group of citizens in San Diego created an association organized to "hunt" clandestine migrants attempting to cross the US-Mexico border in the rural areas east of the city. While the group's organizer stated that they would contact immigration officials once illegal migrants were identified, he also confirmed that the organization's members would be armed "for their own protection".

This anecdote illustrates how border areas are those most exposed to international phenomena and how they are the places where their impacts are felt most strongly. By examining developments in these regions, one can effectively study how international phenomena affect local communities and, by doing so, identify what mechanisms can be implemented to effectively manage these pressures. By comparing models of supranational integration and cross-border cooperation (or lack thereof) at the sub-national level, this chapter attempts to identify sources of power, defined as the ability to manage national borders.

Current immigration policies in Europe and North America have proven to be ineffective and dangerous. Not only have they failed to keep would-be migrants out of developed states (see part four), they have also cut off the safest and most economically beneficial forms of migration (commuter migration) while at the same time feeding clandestine immigration systems. As stated earlier, this has made the migration experience more dangerous and increased the importance of criminal influences. It has also created a negative atmosphere around migration which has led to nativist reactions, such as the one described above.

The San Diego-Tijuana case presented above illustrates the perverse effects of border control strategies and parallel economic development. The physical, political and economic walls that have been constructed between these two cities, and others along the US-Mexico border, have actually stimulated clandestine migration in the area, which continues without regulation. The number of clandestine migrants intercepted at the southwest US-Mexico border increased every year from 1993 (979,000) to 2001 (1.6 million). After a decline in 2002 and 2003, the number of apprehensions has risen again in 2004 to 1.16 million.[38] Scholars of migration flows in the area, most notably Wayne Cornelius, have argued that the recent decrease in annual interceptions is more related to the improved ability of smugglers to avoid apprehension than it is to deterrence. This interpretation is supported by the increase in

---

[38] Cornelius, W. 2004. "Controlling 'Unwanted' Immigration: Lessons from the United States, 1993-2004". Working Paper 92. San Diego, CA: Center for Comparative Immigration Studies, Dec., p. 30.

internal apprehensions in Southern California, (most notably in San Diego, Los Angeles and Orange County), the growth in the number of immigrant deaths along the US-Mexico border, and the escalation of violence involving competing networks of smugglers, which has resulted in murderous attacks on migrants in California, Arizona and New Mexico. The governors of each of these states has recently declared states of emergency in their border counties as the amount of border crime and the number of migrant deaths in these areas has reached record levels. This would seem to indicate that unregulated market expansion encouraged by NAFTA, and the resulting social marginalization found in binational communities, such as San Diego-Tijuana (both between and within the case cities) have stimulated crime and unregulated migration flows in the area. Scholars of case regions in Europe characterized by similar political and economic relationships, such as parts of the Spanish-Moroccan and German-Polish borders, have noted similar outcomes (see Andreas, Scott).

This conclusion is also supported by recent ethnographic work with clandestine migrants in the San Diego-Tijuana region. According to Tito Alegria, immigration controls have prevented legal commuter migration but they have not deterred clandestine migration at all. His research indicates that coercion has had little impact on labor decisions made by migrants in the area. Of the transnational migrants that he interviewed, 53% did not have legal papers.[39] Moreover, his project showed that Tijuana, itself has become the target for educated and specialized migrants due to its recent economic expansion. These internal migrants, in turn, have expressed little overt interest in crossing the border, and they usually remain in Mexico. Alegria's study shows that most of those immigrants entering the US clandestinely are those who cross the border because they are caught outside of Tijuana's regular labor market or earning minimum wage, marginalized socio-economically, and looking for work in San Diego's informal economy.

Conversely, the key to immigration management in border areas seems to be tied to binational development, symmetrical cross-border cooperation between sub-national officials and socio-economic equity. In Bari-Durress, where border integration has occurred, the impact of political and economic development on migration trends is clear. During the early 1990s, between 10,000 and 20,000 clandestine Albanian migrants were intercepted by the Italian coast guard in the province of Bari every year. The high point of these flows was 1996 when 21,000 clandestine migrants were caught along local shores. Since then, clan-

---

[39] Alegria, T. 2002. "Demand and Supply of Mexican Cross-Border Workers". *Journal of Borderland Studies* 17, 1 (Spring), p. 40-1.

destine migration to Bari has dropped off significantly, falling to under 10,000 interceptions annually.[40] Moreover, as smuggling routes were pushed further south, the nationalities of those immigrants intercepted significantly changed from Albanians to Asians and North Africans, signifying a shift in Albania from a sending country to a transit state. The number of migrant deaths in the Adriatic Sea has also decreased significantly. This seems to suggest that the improved institutional and economic situation in Albania, combined with border patrols, has discouraged systemic clandestine migration in the Bari-Durres area. According to one local police official: "The trafficking rafts full of clandestine migrants don't arrive in Puglia anymore, not since a couple of years ago. Now we find a few kids hidden in the trucks coming from Albania, Greece, and Turkey. They are groups of two or three people, usually relatives.[41]" Furthermore, because of the cooperation between the city of Bari and various partners in the Balkans, there is little ethnic chauvinism in the local population. Instead of "hunting" clandestine migrants like their counterparts in San Diego, the inhabitants of Puglia, where Bari is located, have been applauded for their reactions to this phenomenon and they were even nominated for the Nobel Peace Prize by the Italian government.

Regional integration in the Adriatic Basin has not only positively affected border security through development, but it has also brought together authorities from different levels of government who collaborate institutionally to combat crime in the area. By involving sub-national actors more significantly in institutionalized cross-border security discussions, more coherent strategies can be developed that link economic development to immigration management and the reduction of criminal influences. These phenomena are both transnational and interconnected by nature so the development of policies that recognize these relationships and extend beyond borders is fundamental. Sub-national actors can become an integral part of this process. For example, in August 2005 the governors of the US state of New Mexico and the Mexican state of Chihuahua agreed to raze the small town of Las Chepas, a strategic border point for human smugglers, commit state funds to border security and share local law enforcement information. More significantly, they renewed discussions concerning the development of the binational border region.

Unfortunately, the use of soft power is currently being ignored by European and American leaders in the creation of migration control

---

[40] Personal Interview with Italian Coast Guard Official conducted 10 October 2001.

[41] Reported in Zagaria, C. 2006. "Bari, tre clandestini morti in un container". *La Repubblica* 7 marzo, p. 23.

strategies. Soft power focuses on the socio-economic context that surrounds migration politics and it provides a coherent approach to migration management within the framework of regional development strategies promoted by the EU and NAFTA. Most studies of border communities, including this one, in fact suggest that these areas are characterized by inter-institutional conflict as sub-national leaders often follow economic development incentives that contradict national security strategies. Moreover, as the above-mentioned recent developments in the American southwest demonstrate, local communities often bare the financial and political costs of ineffective border control strategies based on coercion and deterrence.

This study demonstrates that national policies do not dictate development in border communities. Instead, the cases presented in this study indicate that contemporary developments closely reflect the interaction of local interests and supranational models of regional integration. This integration, as well as economic globalization, in general, is a phenomenon that worries many native citizens in developed states, who in turn pressure their representatives to invest human and financial resources in expensive border controls. However, this chapter shows that short-sighted and unbalanced development strategies in border areas, rather than regional integration, are responsible for challenges to border security in these communities, such as clandestine migration. Recently, the Mayor of the US border city, Douglas, Arizona, stated, "I have seen illegal immigration all my life. Illegal immigration has a life of its own. You can't stop it[42]". Instead, of continually trying to do so, it seems time to change course and address the underlying causes of this phenomenon. The evidence from this comparative study of border communities would suggest that cross-border development contributes to the reinforcement of border security both structurally and economically. Thus, accelerating border integration would actually by the most effective means of border control.

References

Albanian Economic Development Agency. Website. http://aeda.gov.al/ achievements.htm [accessed June 2005].

Alegria, Tito. 2002. "Demand and Supply of Mexican Cross-Border Workers". *Journal of Borderland Studies* Vol. 17, 1 (Spring), p. 37-55.

Alarcon, Rafael. 2005. "Mexican Migration Flows in Tijuana-San Diego in a context of Economic Uncertainty", in Kiy, Richard and Woodruff, Christo-

---

[42] Reported in Vartabedian, R. 2005. "In a State of Emergency, City's Relaxed". *Los Angeles Times*, 25 August.

pher. eds. *The Ties that Bind Us: Mexican Immigrants in San Diego County.* La Jolla: Center for US-Mexican Studies, UCSD.

Amendola, Giandomenico. 1985. *Segni & Evidenze.* Bari: Edizioni Dedalo.

Andreas, Peter. 2000. *Border Games.* Ithaca: Cornell University Press.

Anguiano-Tellez, Maria Eugenia. 2005. "Cross-Border Interactions: Population and Labour Market in Tijuana", in Kiy, Richard and Woodruff, Christopher. eds. *The Ties that Bind Us: Mexican Immigrants in San Diego County.* La Jolla: Center for US-Mexican Studies, UCSD.

Appendini, Kirsten A. and Bislev, Sven, eds. 1999. *Economic Integration in NAFTA and the EU: Deficient Institutionality.* New York: St. Martin's Press.

Aykac, Aygen. 1994. *Transborder regionalisation: an analysis of transborder cooperation structures in Western Europe within the context of European integration and decentralisation towards regional and local governments.* Libertas.

Baldassare, Mark. 2000. *California in the New Millenium.* Berkeley: University of California Press.

Blatter, Joachim. 2004. "'From Spaces of Place' to 'Spaces of Flows'? Territorial and Functional Governance in Cross-Border Regions in Europe and North America". *International Journal of Urban and Regional Research.* Vol. 28, 3 (Sept.), p. 530-48.

Brooks, David and Fox, Jonathan. eds. 2002. *Cross-Border Dialogues.* La Jolla: Center for US-Mexican Studies, UCSD.

Castells, Manuel and Portes, Alejandro. 1989. "World Underneath: the Origins, Dynamics, Effects of the Informal Economy", in Portes, Alejandro. ed. *The Informal Economy.* Baltimore: The Johns Hopkins University Press.

Chambers, Edward J. and Smith Peter H. eds. 2002. *NAFTA in the New Millenium.* La Jolla: Center for US-Mexican Studies, UCSD.

Chavez, Leo R. 1998. *Shadowed Lives: Undocumented Immigrants in American Society.* Toronto: Thompson Learning, Inc.

Chiquiar, Daniel and Hanson, Gordon H. 2002. *International Migration, Self-Selection, and the Distribution of Wages: Evidence from Mexico and the United States.* Working Paper 9242 Cambridge, MA: National Bureau of Economic Research (October).

City Government of Durres. Website. http://www.tavolovirtuale.tno.it/durazzo.asp. [accessed June 2005].

Clarkson, Stephen. 2000. *Apples and Oranges: Prospects for the Comparative Analysis of the EU and NAFTA as Continental System.* Robert Schuman Centre Working Paper 2000/23. Florence: European University Institute.

Cornelius, Wayne *et al.* eds. 2004. *Controlling Immigration.* Stanford, CA: Stanford University Press.

Cornelius, Wayne. 2004. *Controlling 'Unwanted' Immigration: Lessons from the United States, 1993-2004.* Working Paper 92, San Diego, CA: Center for Comparative Immigration Studies (December).

Di Comite, Luigi and Valleri, Marisa A. eds. 1994. *Problemi demo-economici dell'Albania.* Lecce: Argo.

Erie, Steven P. 1999. *Toward a Trade Infrastructure Strategy for the San Diego/Tijuana Region*. Briefing Paper, San Diego Association of Governments, (February).

European Commission. 2002. External Relations Directorate General. *Albania: Country Strategy Paper: 2002-2006.*

European Commission Delegation to the Republic of Albania. *Assistance.* www.delalb.cec.eu.int/ep/eu, 2004 [accessed July 2005]

Gathmann, Christina. 2004. *The Effects of Enforcement on Illegal Markets: Evidence from Migrant Smuggling along the Southwestern Border.* IZA Discussion Paper 1004 (January).

Guardia, Teresa (Ministry of the Interior, Spain). 2002. *New Strategies in the Fight Against Trafficking in Human Beings During the Spanish Presidency of 2002.* Madrid.

Hanson, Gordon H. 2003. *What Has Happened to Wages in Mexico Since NAFTA? Implications for Hemispheric Free Trade.* Working Paper 9563 Cambridge, MA: National Bureau of Economic Research (March).

ISTAT. 1996. *Puglia: Informazioni Utili.* Rome: ISTAT.

Jamieson, Alison and Silj Alessandro. 1998. *Migration and Criminality: the Case of Albanians in Italy.* Working Paper 1 Rome: Ethnobarometer.

Joseph, Gilbert M. and Henderson, Timothy J., eds. 2002. *The Mexico Reader.* Durham, NC: Duke University Press.

Kirchner, Emil. 1998. *Transnational Border Cooperation Between Germany and the Czech Republic: Implications for Decentralisation and European Integration.* Robert Schuman Centre Working Paper 98/50. Florence: European University Institute.

Kiy, Richard and Woodruff, Christopher, eds. 2005. *The Ties That Bind Us: Mexican Immigrants in San Diego County.* La Jolla, CA: Center for US-Mexican Studies, UCSD.

Laursen, Finn. 2003. *Comparative Regional Integration: Theoretical Perspectives.* Aldershot: Ashgate.

Maganda, Carmen. 2005. "Collateral Damage: How the San Diego-Imperial Valley Water Agreement Affects the Mexican Side of the Border", presented at workshop: *Regions, Borders, and States: Comparative Analysis of Contemporary Development sin Sub-national and Transnational Areas.* Florence: Robert Schuman Centre for Advanced Studies, European University Institute, May 2005.

Martinez, Oscar J. 1996. *US-Mexico Borderlands: Historical and Contemporary Perspectives.* Wilmington, DE: Jaguar Books.

Mattli, Walter. 1999. *The Logic of Regional Integration.* New York: Cambridge University Press.

Morone, Gianni, ed. 1999. *Sviluppo Umano e Sostenibile in Albania.* Milano: FrancoAngeli.

Newspaper Articles were taken from *The San Diego Union Tribune, La Frontera* (Tijuana), *La Gazzetta del Mezzogiorno* (Bari) and *La Voix du Nord.*

Nye, Joseph S. 2004. *Soft Power*. New York: Public Affairs.

Operation Gatekeeper Factsheet. www.crlaf.org/factsht.htm. [accessed April 2005]

Papademetriou, Demetrios G. and Meyers, Deborah Waller, eds. 2001. *Caught in the Middle: Border Communities in the Era of Globalization*. Washington, D.C.: Carnegie Endowment for International Peace.

Pastor, Robert A. 2002. "A Regional Development Policy for North America: Adapting the European Union Model", in Chambers, Edward J. and Smith, Peter H. eds. *NAFTA in the New Millenium*. La Jolla: Center for US-Mexican Studies.

San Diego Association of Governments. 2004. *Census 2000 Fact Sheet*.

Schirm, Stefan, A. 2002. *Globalization and the new regionalism: global markets, domestic politics and regional cooperation*. Cambridge: Polity Press.

Scott, James. 2005. "Cross Border Regionalisation in an Enlarging EU: Hungarian-Austrian and German-Polish Cases", presented at workshop: *Regions, Borders, and States: Comparative Analysis of Contemporary Development sin Sub-national and Transnational Areas*. Florence: Robert Schuman Centre for Advanced Studies, European University Institute, May 2005.

Tijuana Economic Development Corporation. City of Tijuana. Website. http://www.tijuana-edc.com/deitac/ [accessed March 2005].

Viesti, Gianfranco. 1998. *Bari: Economia di Una Città*. Roma: Laterza.

Weiler, J.H.H. ed. 2000. *The EU, the WTO, and NAFTA: Towards a Common Law of International Trade*. Oxford: Oxford University Press.

Weintraub, Sidney. 1992. "North American Free Trade and the European Situation Compared". *International Migration Review* Vol. 26, 2 (Summer), p. 506-524.

World Bank Group. 2005. *Albania at a Glance*.

World Bank Group. 2005. *Albania Data Profile*.

World Bank Group. 2005. *Mexico at a Glance*.

World Bank Group. 2005. *Mexico Data Profile*.

World Bank Group. 2005. *World Bank Assesses Albania's Economy, Outlines Reform Challenges*.

CHAPTER 7

The Border Bottleneck

Drug Trafficking and Incentives for Police Corruption

Daniel SABET

I. Introduction

While the acceptance of free trade between the United States and Mexico has led to a breaking down of the border, the political division between the two countries remains a reality. No example is more para-digmatic of this fact (with life and death implications) than drug traf-ficking. Mexico has the unfortunate fate of being located next door to a large demand for drugs in the United States. Drugs flow relatively easily northward through Mexico; however, the border creates a bottleneck as supply attempts to meet demand. Filtering supply through this bottle-neck has led to the establishment of Mexico's major drug cartels all along the border.

The main tool to combat drug trafficking is law enforcement agen-cies. Unfortunately, however, the professional development of law enforcement in Mexico has traditionally been neglected (Lopez Portillo 2002), despite, or perhaps as a result of, the considerable incentives for corruption, malfeasance, and political manipulation. The bottleneck created by the border, and the profitability presented by the drug trade, increase the incentives for corruption and the desirability of corrupting law enforcement officials. As a result, the criminal justice system, rather than becoming part of the solution to drug trafficking, converts itself into part of the problem.

This chapter first documents the reality of the border bottleneck by examining recent violence in Northern Mexico. The chapter then uses a political economy approach to explain the incentives for corruption and demonstrate how these incentives are altered by the presence of organ-ized crime. Approaches to combating corruption by municipal, state, and federal entities are then documented. Although reforms can have a significant impact (and greater reforms could have a much stronger impact), the fundamental reality of the border bottleneck creates strong incentives for corrupt behavior that public policy tools can only partially address. A long term solution to the problem of corruption will require breaking citizen cynicism towards law enforcement and active participation in holding public officials accountable.

II. Rules and the Border

According to the US-Mexico Chamber of Commerce, the United States and Mexico conducted 235.46 billion dollars worth of trade in 2003. For many, heavy volumes of trade are evidence that the political border between the US and Mexico has less relevance as supply meets demand without the infringement of unnecessary tariffs. Nonetheless, the border has not disappeared entirely in the economic policy arena. Businesses wishing to move goods from south to north, still must confront long lines, submit to revisions, provide documentation regarding the origins of their products, and meet a number of other requirements. There is therefore a bottleneck at the border, and numerous businesses have emerged to facilitate the flow of goods and negotiate the different rules across the political division.

Despite the North American Free Trade Agreement, certain goods are still prevented easy access across the border. For example, the US has attempted to block the sale of avocados and Mexico the sale of corn syrup. Rules restricting the flow of these goods have made the border an even tighter bottleneck, with higher prices on the other end. However, the economic goods most heavily impacted by the border are illegal drugs. Rules that make the drug trade illegal and efforts to restrict the entry of such goods into the US market, constrict and lengthen the bottleneck at the border. This does not mean that goods are unable to get to market. Despite strong enforcement efforts, the rules making drugs illegal are insufficient to overcome the powerful incentives created by large profits to be earned from meeting high inelastic demand in the United States. Rather, the illegality of drugs creates a very tight bottleneck and a very high price for the cartels that are able to bring supply of an illegal good to markets across the border.

The transportation of drugs across the western part of the border has been dominated by the Arellano Felix brothers' cartel. After the death of his brother Amado in 1997, Vicente Carrillo Fuentes took up the leadership of the Juarez cartel, which has traditionally dominated the central region of the border and is considered to be aligned with Joaquin "El Chapo" Guzman of the Sinaloa cartel. The east has been dominated by the Gulf Cartel, which is believed to be led by Osiel Cardenas, despite his arrest in 2003.

Because of inter-cartel violence and government efforts to fight the drug cartels, 2004 and 2005 have been particularly bloody years in Mexico. From January to August 2005, media reports placed the number of dead from drug related violence at 820 (El Universal 2005). No place has typified the rising insecurity better than Nuevo Laredo, Tamaulipas. The city gained international media attention when the head of the municipal police was assassinated less than 24 hours after he took office announcing that he was beholden to nobody (Marshal 2005). But drug violence has occurred throughout northern Mexico and in other key locations such as Acapulco. In Sonora, for example, convicted drug trafficker Efraín Beltrán Félix was sprung from a Sonoran hospital by 15 armed men (Nevarez 2005). In Chihuahua, officials discovered eleven dead bodies buried in a backyard in a residential neighborhood (El Diario 2004). Throughout the region, assassinations and kidnappings regularly fill the local newspapers.

III. Law Enforcement and Drug Violence

The state's security apparatus and law enforcement agencies are the primary mechanism to combat the drug cartels. Law enforcement in Mexico is divided by preventive and investigative functions. Preventive police conduct patrols, interact with the community, provide security, and are the first to respond to criminal acts. Preventive police make up the largest portion of policing and operate primarily at the municipal level, but also at the state and federal levels. Investigation of criminal acts, on the other hand, is carried out by investigative police, known as judicial or ministerial police, which operate at the state and federal levels.[1]

The drug trade is considered a federal crime; and therefore, it falls under the jurisdiction of federal authorities. Nonetheless, all levels of policing are affected by the drug trade. Powerful drug traffickers, with large resources budgeted for bribery and a willingness to use violence to coerce, have led to co-opted enforcement. Frequently police officers and departments look the other way to drug activity, and, in the worse case scenario, they are actively involved in the drug trade.

Stories of police corruption along the US-Mexico border are as shocking as those of drug violence. In 1997 the head of the National Institute to Combat Drugs (INCD) Jesús Gutierrez Rebollo was arrested and charged with collaborating with the Juarez cartel. In the 1990's, deserters from an elite anti-narcotic unit in the Mexican army formed

---

[1] See Lopez Portillo (2002) or Reames (2003) for an in-depth description of the Mexican police.

the Zetas, a heavily armed group considered to be the enforcers of the Gulf Cartel. Papers filed by the government of Mexico in San Diego in 1997 revealed that the Baja California state attorney and a large percentage of law enforcement officials and judges had been compromised by the Arellano-Felix Cartel (Constantine 1997). More recently, in the above mentioned Nuevo Laredo case, the municipal police were considered to be so inter-connected with the drug trade that the federal government temporarily disbanded and disarmed the entire force (Marshall 2005).

The costs of such corruption are significant. At the very least, police corruption leads to the impunity of the drug cartels. In reality, however, the impacts are much greater. With no one to police the police, the result is a failed justice system. Using 1998 data from the Secretary of the Interior, Bailey and Chabat (2002) find that of 1.49 million crimes reported (a fraction of the actual number of crimes) and 1.33 million preliminary inquires initiated, only 149,000 arrest warrants were issued and only 85,000 were actually carried out. The failure to realize justice sets in motion a vicious cycle, as it produces a cynical citizenry that distrusts the police, is unwilling to report crimes, and is unable to participate in fixing the problems in the criminal justice system.

IV. Incentives for Corruption

Police, by their very nature, face strong incentives to engage in corruption, defined as the abuse of position and authority for personal benefit. Police corruption is not specific to the border, nor is it only the product of the drug trade; however, the presence of the drug trade raises the benefits to be gained from corruption and increases the cost to not engage in corrupt practices. I argue here that in the presence of an aggressive organized crime syndicate, corrupt behavior is as rational strategy, which when scaled up across a police agency, further increases the incentives for corruption and undermines the rule of law.

Consider the interaction between a police officer and a common drug dealer caught in the act of selling illegal drugs without the protection of a strong organized crime syndicate. The dealer is engaged in illegal activity punishable by law. The officer, on the other hand, has not engaged in illegal activity, is charged with enforcing the law, and is supported by the law. As a result, the officer is in a position to dictate the terms of interaction between the two. The officer could chose to fulfill his duty, arrest the alleged criminal, and follow proper procedure in doing so. Alternatively, the officer could exploit the situation and extort the alleged criminal for a share of the dealer's profits. In such a situation, the alleged criminal would have little choice but to comply

with the terms dictated by the officer. If a corrupt interaction is repeated, the terms of trade shift, for the officer has become engaged in
illegal activity. The interaction comes to represent an exchange that is
mutually beneficial to both actors. The drug dealer is allowed to operate
with protection from prosecution and the police officer receives a
kickback in return.

Now consider the interaction between a police officer and a drug
trafficker given the presence of a strong organized crime syndicate.
While the trafficker is still engaged in an illegal activity and the officer
still has the law on his side, the officer is not necessarily in a position to
dictate the terms of the interaction. An organized crime syndicate
provides its members with leverage and protection in such interactions.
In the previous example, the officer had to choose between fulfilling his
duties and rent seeking. The opportunity cost of fulfilling his duty was
only missed rents. In the second scenario, however, as the crime syndicate offers protection to its members, the opportunity cost of fulfilling
an officer's duty might also include retribution from the syndicate.
Given repeated interactions the crime syndicate's position is further
strengthened by the officer's illegal activity. As in the previous scenario, however, the interaction is still self sustaining as both officer and
criminal benefit from the arrangement.

Unfortunately, what is a rational action for an individual officer has
disastrous collective consequences when scaled-up. When one officer
defects from her duty and cooperates with crime syndicates, the action
is lost within the numerous interactions between officer and society.
However, when numerous individual officers defect from their duty, the
law loses meaning and the judicial system fails. To illustrate, if the
crime syndicate has made significant inroads into the criminal justice
system and is able to ensure that charges are dropped against their
members, then every police officer loses his or her ability to offer a
credible threat against members of organized crime. Fulfilling one's
duty looses its rationality if the conviction of a guilty criminal cannot be
upheld and an officer faces retribution for carrying out an arrest. In this
highly advanced stage of corruption, malfeasant behavior becomes the
dominant strategy for all officers in the force.

The difference in corruption given the presence or absence of organized crime can be seen in different case studies on police corruption.
Consider for example the difference between police corruption in New
York prior to, during, and after prohibition. Sloat (2002) offers a study
of corruption in New York from 1892-1895. He describes an aggressive
police force collecting protection money from businesses, abusing
authority, and seeking rents. In the described interchanges, the police
officers (so long as they were loyal to Tammany Hall) are able to act

with impunity. Lawrence (1974) describes a very similar version of police corruption in the city in the 1960's. He depicts officers planting evidence, illegally using wire taps, and manipulating drug users for personal gain. Stolberg (1995), however, portrays a very different image of police corruption in New York during the prohibition era. The criminalization of alcohol in the face of widespread demand led to strong organized crime activity. Not only did police corruption increase during this time, but criminal elements came to dominate the terms of trade. Stolberg (1995) writes:

> Prohibition proved to be a formidable barrier to good and honest government... The result was what one prosecutor called 'a rapidly growing cancer of public corruption. '[Prohibition] was organizing the underworld as it had never been organized before, and it was endowing the underworld with immense sums of money for corruption'. (p. 16)

The problem is not specific to New York. In fact, a large literature documents the nexus between police corruption, organized crime, and prohibition in the United States (Allsop 1968; Landesco 1968; Chambliss 1971). Allsop (1968) cites statistics from 1923 that estimate that 60% of Chicago's police force was involved in the liquor business. He writes, "During the twenties Chicago was effectively a city without a police force, for it operated partially as a private army for the gangs..." (p. 17).

The presence and absence of organized crime has resonance in the modern world as well. For example, consider Gay's (2005) description of police corruption in the low income areas of Rio de Janeiro. While criminal activity is organized through gangs, they lack the power of a strong drug cartel. In this environment, Brazilian police are able to control the terms of trade in interacting with drug dealers. As Gay (2005) writes, "When the police got a hold of someone, they called the dono [leader] of the drug gang on his cell phone and began to negotiate. The higher up the person on the drug gang ladder, the higher the asking price". (p. 87). By contrast, the dominance of the drug cartels in Colombia during the 1980's and 1990's allowed them to dictate the terms of interaction with the police. The payroll of the Cali Cartel in Colombia obtained during a raid included 35 senior police, 24 mid-level police, and 130 agents (Staff Report 1996).

In Mexico today there is certainly no hesitation on the part of drug traffickers to make good on their threats. In the 1990's the Arrellano-Felix cartel operating out of Tijuana earned the reputation as the most violent organized crime syndicate assassinating two municipal police chiefs in 1994 and 2000, a subdirector of the Tijuana office of the Institute for the Combat of Drugs (INCD), and the director of the federal police force in Tijuana, among many others. The federal director was assassinated just two days after stating that the police had become so corrupt they weren't just friends with the traffickers, they were their servants (cited in Constantine 1997). The recent upsurge in violence in 2005 has caused the assassination of a large number of public security officials including the above mentioned case of Nuevo Laredo.

V. Efforts to Combat Corruption

There are a number of theoretical approaches that address the issue of corruption. The following discussion examines corruption as a problem of bad apples, culture, institutional carrots, institutional sticks, and of a lack of citizen involvement and oversight. Mexican policy makers have at different times and in different locations attempted to address corruption problems from these different angles. However, each approach has lead to actions that were necessary but insufficient to address what has become a complex and endemic problem. Moreover, a powerful organized crime syndicate with a large supply of money and a willingness to use violence to threaten and enforce is able to exploit the weaknesses in any individual approach.

While this section discusses corruption in general terms, the data for this study was obtained primarily through interviews with public security officials at the state and municipal levels and members of civil society in the northern border states of Baja California, Sonora, Sinaloa, and Nuevo Leon. Due to the delicate nature of corruption issues, reference cannot be made to specific police departments. Nonetheless, this following discussion provides an accurate overview of police corruption along the US-Mexico Border.

Bad Apples

The root cause of police corruption is often considered to be discretion and low public visibility (Walker 1993). Bad apples, it is felt, take advantage of low visibility and discretion to engage in rent seeking behaviors. Approaches to remove bad apple reforms have two elements. The first is preventing bad apples from entering the police force, and the second is removing bad apples that are in currently in the force. Traditionally there has been minimal screening of the police. In an interview with Suarez (2005), one former state police officer in prison for participating in drug trafficking stated "The institution and the government do not worry much about selecting personnel. In the police they say, "Do you have secondary [up to grade 9]?", "Yes", "Come on, because we lack police".[2]

One reform that many municipal and state police agencies have undertaken in recent years has been to make a high school degree a requirement to enter the police. On average, however, most police have no more than a secondary education and a great number of those have less education (Carrasco Araizaga 2003).The police departments that do not require a high school degree fear that if they raise the education requirement, they will not be able to find a sufficient number of recruits. The fact that numerous police departments now require a high school education and have not had a shortage of recruits would suggest that this fear is unsubstantiated, at least in the urban areas. In addition to education requirements, many departments have instituted detailed psychological testing of applicants to the police academy.

The other side of addressing the bad apple problem is removing spoiled apples from the force. As will be discussed below, the institutional mechanisms responsible for following citizen complaints concerning corruption are week, and bad apples cannot be removed if they cannot be identified. One mechanism that is growing in popularity in Mexican policing is the use of random drug tests. While a corruption investigation is a subjective process vulnerable to the vicissitudes of internal and local politics, drug testing represents an easily measurable, relatively clear cut method to identify potential bad apples. However, hiring someone with a higher education level and firing someone that is using illegal drugs are only very indirect means to address the problem of malfeasance. Someone with a low education level can still be very professional and someone that avoids drug use can still be highly involved with drug traffickers.

More importantly, while removing bad apples is necessary, complete reliance on such approaches assumes that corruption is a disease of the morally weak, and it fails to appreciate the rationality of corruption. In fact, a number of scholars have documented the process through which good cadets turn into corrupt officers over time (Sherman 1974; Steffens 1931; Suarez 2005). These studies point out that decisions to engage in corrupt activities are not isolated choices, rather they are both embedded in and dependent upon the actions of other officers and corrupting influences. The presence of corruption in a hierarchical police department means that new recruits face strong internal pressure to engage in malfeasant behavior. The presence of drug traffickers

---

[2]   In Spanish the quote reads, "La institución y el gobierno no se preocupan mucho por seleccionar al personal. En la policía dicen: "¿Tienes secundaria?", "Sí", "Vente porque hacen falta policías".

means that there are individuals and groups actively attempting to corrupt police officers.

Police Culture

Cultural approaches to corruption recognize the embedded nature of decisions to engage in corruption. Police operate both within the formal rules of laws and police procedure and the informal rules of their unit and department. The US literature on police subculture has focused on how the informal rule of "secrecy" to protect fellow officers and strengthen internal solidarity has created an environment conducive to police misconduct (Stoddard 1968). Informal rules might be even more blatantly insidious, however. In some cases, a bribe is required to enter the police force (Alfiler 1986; Sherman 1974) and once on the job police might be required to collect a quota of bribe money to be passed on to higher level officials (Rose Ackerman 1999, p. 81). In such contexts, the informal rules of policing create very strong incentives to engage in malfeasance both to avoid punishment from superiors and to achieve promotions or good assignments. As a result, the decision to engage in malfeasant behavior cannot be divorced from the embedded nature of police decisions within a police culture of informal rules and within a hierarchy where informal rules can be enforced.

Mexican anti-corruption reformers are well aware of the existence of informal rules in the police department and have at times sought to address it by recreating the police. For example, in one municipality with three policing districts, the existing force was distributed into two of the three districts and new cadets were all assigned to the third district. At the federal level in 1999 the preventive police were reorganized into a new Federal Preventive Police force with strict parameters for professional development and training (Lopez Portillo 2002). In a number of states, state governments have attempted to create new elite preventive police forces.

Such approaches suffer from two problems. First, they fail to acknowledge the realities of immediate manpower needs. A new police department cannot emerge from scratch. Lopez Portillo (2002) points out that all the 100,000 Federal Preventive Police were merely brought in from the military, the federal highway police, or the Center for Research and National Security (CISEN). More importantly, however, while such approaches recognize the realties of informal rules, they fail to acknowledge the way in which informal rules are created. In New York in the 1960's, investigations into corruption in the city's Narcotics Division led to a gradual transfer of almost the entire staff of the division. Three years later a study found continued corruption. Over the course of the following years almost 100% of the division was transferred out; however the problem of corruption remained (Sherman 1974). As the initial theoretical discussion of this chapter argues, there are strong individual incentives to engage in corrupt behavior, and, given the presence of organized crime, there are powerful agents that are actively seeking to corrupt police officers. While informal rules increase the incentives for malfeasance and make it difficult to fight, their removal does not eliminate the underlying causes of corruption.

One method to address both the cultural and individual incentives for corruption is through professionalization. Professionalism creates a positive culture that informally rewards good behavior and discourages corruption. Unfortunately, efforts at professionalization often only go as far as slogans, ethics courses, and new uniforms. Neiderhoffer (1969) writes that ironically, one of the initial sources of cynicism in the New York police was the strong emphasis placed on professionalism and ethics in the academy that simply did not match with reality. While preparation in values, ethics, and the need to be proponents of the rule of law is important, training must embrace the realities of police work and not build an unrealistic schema that shatters during the first week on the job. As the next section argues, professionalization means that policing becomes a viable career choice where the incentive to take a bribe is less than the incentive to perform one's duty.

Salary and Job Mobility

Corruption is inherently an economic transaction and material gain is frequently the reward for corrupt officials. Salaries of law enforcement officials are notoriously low, causing some to conclude that bribe taking is an informal part of an officer's salary. Preventive municipal police forces in northern Mexican cities typically pay around 5,000 pesos a month (US$460) to entry level officers. A few municipalities have attempted to change this status quo, and now pay as much as 8,600 pesos (US$792) a month. In addition, as state governments attempt to develop elite preventive forces (which will be discussed further below), they frequently earn a similar amount. Judicial or ministerial police also earn a higher salary. When wages that are insufficient to gain entry into the middle class are combined with the considerable opportunity for corruption and a large supply of drug trade created bribe money, corruption appears almost inevitable. Nonetheless, few governments have been willing to absorb the high costs of across the board salary increases, and interview respondents frequently cited limitations in public resources in their jurisdiction. In a recent symposium on police reform, it was argued that in Mexico City 70% of the public security budget already goes to payroll (Center for Strategic and International Studies 2004).

Nonetheless, an increased salary has two important effects. First it reduces the ability to justify a bribe both in the mind of the individual officer and in front of society more generally. Second, it reduces the value of an actual bribe. This is to say that a bribe of US $20 has greater value to an officer that earns US $460 a month than to a police that earns US $792 a month. Unfortunately, however, given that drug traffickers depend on corrupt officers to carry out their business, an increase in salary will not eliminate efforts to corrupt officials. A raise in salary might only lead to an increase in the bribes that drug traffickers must offer police. In fact, corruption often occurs at high levels of the police hierarchy, where public employees gain high enough salaries to belong to the middle or even upper class (Rose-Ackerman 1999). As a result, some scholars contend that at least some portion of rewards should be based on performance (Rose Ackerman 1978). In several departments, officers can see their salary close to double if they do not miss work, arrive on time, and do not receive any faults on their record.[3]

An even stronger incentive to avoid corrupt activities is a realistic opportunity to move up in the police hierarchy. As Sherman (1974) writes, "The greater the policemen's perception of legitimate advancement opportunities, the less likelihood there will be of their accepting corruption opportunities (p. 22). For many officers with a low discount rate (the ability to understand their own self interest in the long term) a bribe is not worth sacrificing a promising career. Unfortunately, however, policing in Mexico is not a career with upward mobility. Interview respondents were quite frank about the fact that promotion is based on personal ties rather than merit. Respondents reported that it is not uncommon for an officer with ten years in the force to be making almost the same salary that he or she started with. Civil service reform is regularly raised as a means to turn policing into a career; however, no such reform has been forthcoming at the state and local level. While civil service reform could offer a means to ensure advancement, in its current absence, meritocratic promotion practices with clear criteria for advancement could still be instituted. One border municipality has instituted a program whereby supervisory officers and aspirants must reapply or apply annually for leadership positions. A committee that includes civil society representatives reviews the applications. In addition to many other criteria, the committee conducts economic background checks to ensure that officers' assets are not disproportionate to their earnings.

---

[3] It should be noted that such efforts to reward police performance should not place undue amounts of decision making authority in the hands of supervisors who they themselves might be corrupt.

While clearly there is an urgent need to increase the financial incentives police face to faithfully fulfill their duty in accordance with the rule of law, improved pay and upward mobility cannot alone solve the problem of corruption. As Rose-Ackerman (1999) writes, "Many public and private agents are responsible for making decisions with financial consequences that far exceed their pay levels. In such cases it is unrealistic to suppose that incentive bonuses can equal a high proportion of the value of the benefit dispensed". (p. 79). This is certainly the case when the drug trade and billons of dollars of profit are at stake.

Avoiding Opportunities

As corruption has its roots in discretion and low visibility, many scholars advocate that reform efforts should seek to reduce these two factors. Terris (1967, p. 145), for example, argues that moving police into police cruisers and off a walking beat in the US reduced the opportunities to engage in corrupt interactions.

Mexican authorities have often favored reducing the opportunities for corruption. Some departments randomly assign officers and constantly move their geographical area of patrol (Lopez-Montiel 2000). If a drug trafficker wants to ensure that officers look the other way during the transport of a drug shipment, random assignment means that he or she cannot be sure who to bribe. In addition, some police patrol in large groups to increase their visibility. Unfortunately, these methods come at a cost. Police never develop a relationship with the citizens that they are protecting, gain the trust of the community, or develop local knowledge about crime problems. As a result police are less effective and less accountable to citizens. In addition, such methods do not necessarily avoid the problem of corruption; in fact, they might only mean that drug traffickers have to bribe officials higher up the ranks to impact how agents are assigned and where they are deployed (Rose-Ackerman 1999).

VI. Institutional Sticks: Who Polices the Police?

Through institutional reforms, police departments are also able to increase the costs of malfeasance (Rose-Ackerman 1999). Police will be less likely to engage in corrupt transactions if there is a high probability of being caught, and if being caught implies alienation, removal from the force, or criminal charges. The presence of sticks first requires strong legal tools criminalizing corrupt behavior. For example, to confront the prevalence of hierarchical corruption schemes where supervisors require a quota of bribe money from subordinates, one border municipality passed legislation stipulating removal from the force for any officer found to be requesting or receiving money from a subordinate or any officer offering money to a superior.[4]

While legal reforms are necessary, the presence of a law does not ensure its enforcement. In fact, in interviews, Mexican officials repeatedly cite the difficulty in implementing existing laws. Beyond criminalizing corrupt behavior, police departments require mechanisms to monitor, investigate, and take action against abuses of authority. Such mechanisms require a strong mandate, personnel, and anonymity and institutional protection for whistleblowers. The most salient distinction between different oversight mechanisms is whether they are housed within the public security agency our outside, frequently with citizen participation. Both proposed solutions suffer from flaws as internal mechanisms are often insufficiently aggressive and external mechanisms run the risk of lowering morale and tarnishing the image of the police.

Unfortunately, however, such mechanisms are perhaps the least developed anti-corruption options in northern Mexico's police forces, and Varenik (2005) argues that there is a need to develop greater accountability mechanisms in Mexico more generally. Accountability mechanisms are not without precedent in Mexico. In the aftermath of a number of scandals including the discovery of ties between the head of the National Institute to Combat Drugs (INCD) Jesús Gutierrez Rebollo and the Juarez Cartel, the federal government created and used a diverse set of accountability mechanisms to clean federal public security agencies. In 1999, through an ethics center (Centro de Control de Confianza), an internal affairs unit, and external review board, 1,517 employees were dismissed and 351 criminal charges were brought (Center for Strategic and International Studies 2004).

Once again, the problem is not so much the lack of accountability mechanisms but problems in implementation. Even at the federal level, accountability mechanisms have declined, and the PGR has recently been questioned about its ability to carry out internal investigation in a number of cases, including a case against two Federal Investigation Agency officers for their potential involvement in the extortion and death of Enrique Salinas, the brother of former president Carlos Salinas de Gortari (Gutierrez 2005). To offer another example, in several municipal police forces there is a Commission of Honor and Justice (Comisión de Honor y Justicia) that is responsible for investigating complaints against officers. To give the commission weight it is composed

---

[4] Some would argue that subordinates should actually suffer a lesser punishment than their superiors to not discourage agents that have had money successfully extracted from them in the past.

of leaders in the municipal government and the department. However, because of their many responsibilities, commission members rarely meet, they are not able to practically carry out investigations, and they leave their mandate unfulfilled. Again, interview respondents were surprisingly frank about the ineffectiveness of existing commissions that have multiple responsibilities, lack sufficient administrative staff, and are led by individuals with more pressing operational obligations.

External oversight commissions involving citizen participation are uncommon. It is frequent that security agencies have committees for citizen participation; however, public participation committees in Mexico typically have administrative objectives rather than oversight capabilities (Sabet 2005). Carrillo (2003) documents one such state public security council. The council was charged with a number of functions including developing legal proposals, revizing monthly re-ports, evaluating the state plan for public security, proposing actions to prevent crime, promoting citizen participation, as well as receiving and investigating complaints about the police. While the council was con-sidered a success and it carried out a number of activities, including as a diagnostic of law enforcement capabilities, a citizen evaluation of law enforcement, crime prevention programs, and text of a law to promote citizen committees, the council's broad mandate meant that its ability to investigate citizen complaints was very limited.

Holistic Responses

As this discussion has illustrated, all of the different reform methods proposed to address law enforcement corruption along the US-Mexico border suffer from flaws. The problem is deeper than bad apples; cul-tural approaches ignore incentives for corruption; and institutional mechanisms can often be circumvented or they produce unintended consequences. Even greater civil society involvement runs the risk of merely serving administrative functions.

Most all of these approaches are necessary but insufficient to address what is an endemic and complex problem. Nonetheless, different reform efforts are not necessarily complementary. For example, mechanisms that randomly rotate police prevent the development of community ties that might create greater accountability. In addition, many efforts are dependent on how they are implemented, not just their presence or absence. For example, the presence of an internal affairs department will not be effective if it is not well constructed, well staffed, and pos-sessing of a strong mandate. There is a considerable propensity for any reform efforts to merely serve as window dressing. Public security agencies are frequently renamed, given new uniforms and new logos, and restructured, but the informal rules continue unaffected at a considerable waste of taxpayers' money.

Mexican authorities at all levels of government have recognized the need for holistic approaches; however, constrained by inertia within existing bureaucracies and facing considerable financial limitations, they often opt to create new elite units rather than address the problems in existing units. Executives with a short time in office seek to start with a clean slate and in the time of their administration cultivate an effective public security force. Such units receive a higher salary, undergo rigorous training, are better equipped, and inculcated with an elite esprit de corps. They are given a great deal of press attention and frequently heralded as a panacea to security problems. The effectiveness of such an approach has yet to be determined, but it certainly fails to address the problems within existing security forces and might even worsen them as all resources, energy, and political capital are directed at the new elite units. Elite forces (especially well publicized ones) also offer drug traffickers a smaller and more focused target in their corrupting efforts. As a result, smaller less politically motivated efforts to reform existing forces and ensure continuity between administrations would probably have a higher long term payoff.

Obstacles

This discussion has touched upon several major obstacles to efforts to address corruption in the context of organized crime along the US-Mexico border. The ability of organized crime to actively circumvent anti-corruption efforts has already been discussed at length. Also raised has been the lack of professionalization and limited opportunities for advancement. Another obstacle that has been hinted at is the problem of leadership. Municipal administrations change every three years and state administrations every six. With the change of administration, all of the top management is frequently replaced. In addition, it is common for the director of the public security agency to change several times during a given administration. This lack of continuity produces a number of problems. First, constant changes in leadership and reforms to the formal rules and procedures within an agency serve to strengthen the informal rules and give them predominance. Second, constant changes in leadership, means that no leader's tenure is long enough to address the endemic problem of corruption. While removing compromised leaders from office is an important means to combat corruption, removing good leaders from office simply because there has been an administration change only prevents the development of effective reforms. Because corruption is a complex problem with many facets, leaders serving a short tenure are frequently only able to focus on one aspect of the corruption problem before they are replaced.

The greatest obstacle to addressing corruption has been the ambivalent role of the citizenry in holding public security officials accountable. For example, assassinations of public security officials often do not incite public outrage and civil society action, for it is often unclear to citizens if an officer was killed because he was doing is job or because he himself was involved in the drug trade. As a result, such assassinations only appear to deepen cynicism and hopelessness rather than produce a civil society response. It is often the case that citizens rationalize the violence in border communities as a fight between drug cartels that does not affect them. Addressing the corruption problem requires that citizens demand accountability from public security officials and create countervailing pressure against such strong incentives for corruption. Reformers would be wise to recognize their own limitations and encourage a more active and even critical civil society.

VII. Conclusion

Different policy arenas are impacted differently by international borders. In the case of the drug trade, the border represents a division between a large supply, a large demand, and huge profits. US law enforcement efforts to prevent the entry of illegal drugs into the United States, creates a bottleneck in northern Mexico as drug cartels attempt to bring their supplies to market. Drug cartels are only able to function effectively in the region by actively corrupting public security personnel, which they are able to do through both the carrots of bribe money and sticks of extortion and violence. The corruption of law enforcement not only allows drug cartels to act with impunity, but it undermines the rule of law, allows for police abuses more broadly, and creates citizen disgust and disengagement from participating in public security efforts.

While there have been a number of government efforts to address the corruption problem, they represent necessary but insufficient changes. A number of obstacles prevent effective reform, the most insidious of which is the active efforts of organized crime to counteract reform proposals. As a result, effectively combating corruption requires a countervailing force of equal strength that can only be found in an outraged citizenry demanding a professional police, reform efforts, continuity in administration, and accountability mechanisms.

References

Alfiler, Ma Concepcion P. 1986. "The Process of Bureaucratic Corruption in Asia: Emerging Patterns", in Carino, Ledivina A. ed. *Bureaucratic Corrup-*

tion in Asica: Causes, Consequences, and Controls. Philippines: JMC Press, p. 15-68.

Allsop, Kenneth. 1968. *The Bootleggers: The Story of Chicago's Prohibition Era.* New Rochelle, NY: Arlington House.

Bailey, John, and Roy Godson. 2001. *Organized Crime and Democratic Governability: Mexico and the US-Mexican Borderlands.*

Bailey, John, and Jorge Chabat. 2002. "Transnational Crime and Public Security: Trends and Issues", in Bailey, John and Jorge Chabat eds. *Transnational Crime and Public Security.* La Jolla: Center for US-Mexican Studies.

Carrasco Araizaga, Jorge. 2003. "Ningún Avance in La Policía Mexicana". *Proceso*, October 12, 34-35.

Carrillo Maza, Marco Antonio. 2003. *La Participación Ciudadana Y El Enfoque Micro Social De La Seguridad Pública: El Caso De Baja California.* La Jolla, CA: Center for US-Mexican Studies.

Center for Strategic and International Studies and the Center for US-Mexican Studies. 2004. "Police Reform". Paper presented at the Justice Reform in Mexico, Washington D.C., July 16.

Chambliss, William J., and Robert B. Seidmann. 1971. *Law, Order, and Power.* Reading: Addison-Wesley.

Sub-Committee on National Security, International Affairs, and Criminal Justice of the Committee on Government Reform and Oversight. 1997. *Counternarcotics Efforts in Mexico and Along the Southwest Border*, February 25.

Diario, El. 2004. "11 Cuerpos Exhumados En Juarez". *El Diario*, January 23.

Gay, Robert. 2005. *Lucia: Testimonies of a Brazilian Drug Dealer's Woman.* Philadelphia: Temple University Press.

Gomez, Francisco. 2005. "Pgr and Dea Alertan Del Poder De Los Arriola". *El Universal*, October 18.

Gutierrez, Alejandro. 2005. "Un Crimen Anunciado". *Proceso*, July 17.

Landesco, John. 1968. *Organized Crime in Chicago: Part Ii of the Illinouis Crime Survey.* 2nd ed. Chicago: University of Chicago Press.

Lopez-Montiel. 2000. "The Military, Political Power, and Police Relations in Mexico City". *Latin American Perspectives* 27, No. 2: 79-94.

López Portillo Vargas, Ernesto. 2002. "The Police in Mexico: Political Functions and Needed Reforms". In *Transnational Crime and Public Security: Challanges to Mexico and the United States*, edited by John Bailey and Jorge Chabat. La Jolla: The Center for US-Mexican Studies.

Marshall, Claire. 2005. "Gang Wars Plague Mexican Drugs Hub". *BBC News*, August 18.

Molina Ruiz, Francisco Javier. 2002. "Organized Crime and Democratic Governability at the US-Mexico Border", in Bailey, John and Roy Godson eds. *Organized Crime and Democratic Governability: Mexico and the US-Mexican Borderlands.* Pittsburgh: University of Pitsburgh Press.

Moore, Molly. 2000. "Mexicans Stunned by Killing of Police Chief". *Washington Post*, February 29.

Nevarez, Oman A. 2005. "Nada Pudieron Hacer Custodios". *El Imparcial*, August 2.

Niederhoffer, Arthur. 1967. *Behind the Shield: The Police in Urban Society*. Garden City, NY: Doubleday.

Reames, Benjamin. 2003. "Police Forces in Mexico: A Profile". Paper presented at the Reforming the Administration of Justice in Mexico, La Jolla.

Rose-Ackerman, Susan. 1978. *Corruption: A Study in Political Economy*. New York: Academic Press, 1978.

———. 1999. *Corruption and Government: Causes Consequences, and Reform*. New York: Cambridge University Press.

Sabet, Daniel. 2005. "Thickening Civil Society: Nonprofits and Problems of Water and Sanitation along Mexico's Northern Border". Dissertation, Indiana University.

Sherman, Lawrence W. 1974. "Becoming Bent: Moral Careers of Corrupt Policemen". In *Police Corruption: A Sociological Perspective*, edited by Lawrence W. Sherman. Garden City, NY: Anchor Press / Doubleday.

Sloat, Warren. 2002. *A Battle for the Soul of New York: Tammany Hall, Police Corruption, Vice, and Reverend Charles Parkhurst's Crusade against Them, 1892-1895*. New York: Cooper Square Press.

Staff Report. 1996. "Corruption and Drugs in Colombia: Democracy at Risk". Washington D.C.: Committee on Foreign Relations.

Steffens, J. Lincoln. 1931. *The Autobiography of Lincoln Steffens*. New York: Harcourt.

Stoddard, Elwyn R. 1968. "The Informal 'Code' of Police Deviancy: A Group Approach to 'Blue-Coat Crime'". *The Journal of Criminal Law, Criminology, and Police Science* 59, No. 2: 201-13.

Stolberg, Mary M. 1995. *Fighting Organized Crime: Politics Justice and the Legacy of Thomas E. Dewey*: Northeastern University Press.

Suárez de Garay, María Eugenia. 2005. "Armados, Enrejados, Desconfiados... Tres Breves Lecturas Sobre La Cultura Policial Mexicana".

Terris, Bruce J. 1967. "The Role of the Police". *Annals of the American Academy of Political and Social Science* 374: 58-69.

Varenik, Robert O. ed. 2005. *Accountability: Sistema Policial De Rendición De Cuentas*. Mexico: Instituto para la Seguridad y la Democracia and the Centro de Investigación y Docencia Económica.

Walker, Samuel. 1993. *Taming the System: The Control of Discretion in Criminal Justice 1950-1990*. New York. Oxford University Press.

Universal, El. "Anti-Drug Operative to Hit Acapulco". 2005. *El Universal*, August 12.

CONCLUSION

Borders, Regions and Cooperation

Michael KEATING

There has been a lot of talk in recent years in Europe and North America about the 'borderless world' or the 'end of territory' as modern communications technology and economic restructuring change the historic relationship between place and function. Less often is the meaning of these terms subjected to critical analysis. Yet as the chapters in this volume have shown, their meanings are by no means obvious, and they vary across time and place.

Borders and boundaries have, since the 19th century, been attributed above all to states or, to use the common but misleading term, nation-states. State and nation building have been historic processes in which the boundaries of polities have been marked and fortified. Within these have developed social systems, cultural communities and the panoply of functions that characterize the modern state. National systems of government became more elaborate in the 19th and 20th centuries as the bureaucratic modern state penetrated peripheral territories, extracting taxes and providing services. Economic systems were moulded to state boundaries within national systems of protection or development. Cultural norms were imposed throughout the territory, through universal education, military service and government-regulated media. The welfare state rested on a complex of nationally-based factors. It provided a counterpoint to national markets, corresponding to the scale of the economic system; it rested upon a conception of social solidarity informed by common membership of a political community and a fixed border determining eligibility; and it corresponded to largely national labor markets. These political, institutional, economic, cultural and social borders therefore largely coincided to sustain each other mutually and underpin the continued territorial integrity of the state. From the 1950s, regional development policies were elaborated within states to incorporate peripheries into national economies in a form of 'spatial Keynesianism'.

This process of incorporation was, of course, never complete. Borders were rarely hermetically sealed and the various political, economic and social systems never fitted perfectly. There were disputed border regions and ethnic and national groups straddling borders. Stateless nations retained a sense of their own identity within and across states. Economic transactions crossed borders and people migrated. Yet something has happened in recent years to change borders, increase their porosity, and delink the various systems that they contained. There has been a proliferation of cross-border initiatives such that there is probably not an inter-state border in Europe or North America without one. One reason for this, perhaps paradoxically, is that borders are no longer challenged politically or militarily. Contested borders tend to create interests around them and state elites are unwilling to countenance cross-border activity for fear of encouraging irredentist forces or weakening the hold of government. Once borders are accepted politically, however, they can be penetrated and by-passed by functional systems. Combined with the broader processes of state restructuring, including transational integration and regional assertion, this favours the delinking of the various political, social, cultural and economic systems previously contained within state boundaries.

Some analyses of the cross-border phenomenon have drawn on neo-functionalist theory to show how the erosion of functional boundaries brings change in the political, social and cultural dimensions of politics. Indeed it is remarkable how neo-functionalism, long superseded in its simple form as an explanation for European integration, comes back as the supposed driver of cross-border cooperation. In fact, the border has now become a site of contestation among multiple forces, some of which tend to favor cross-border integration while others militate against it. The opening of borders benefits some economic actors, who can operate in a larger space and capture new markets, but harms others, used to working in local or national markets. Local politicians may see advantages in cross-border working, where it can achieve improvements in services and infrastructure of benefit to their constituents. They may also gain prestige and publicity for engaging in new forms of paradiplomacy and, in those European countries and regions where the European project is positively regarded, can cement their credentials as good Europeans. On the other hand, cross-border cooperation may bring little immediate reward and, as their constituents are all on one side of the border, they will tend to take a self-interested and often defensive attitude towards common projects. This has often hampered the development of large infrastructures like airports, where one could be of the size to serve the whole cross-border region, but politicians cannot agree on where it should be. Similarly the rationalization of production might mean closures of plants on one side of the border in favor of more efficient ones on the other side. While comparative advantage theory suggests that everyone should gain from the allocation of resources in the most efficient manner, this assumes perfect markets and factor mobility. In practice, gains on one side are not always compensated by equivalent gains on the other. Labor does not always follow capital in relocating across the border or commuting and, even if it does, this will not produce political gains for politicians on the losing side. Cooperation seems to be easier on pure public goods, like environmental improvement and pollution control. State elites in the capital and field services of central governments may also hamper cross-border cooperation, imposing irksome reporting requirements and permissions to ensure that local leaders do not use cooperation to carve out a more autonomous role for themselves.

The opening of borders should not be confused with cross-border cooperation, although the two are sometimes run together. Open borders benefit economic transactions and advantage those actors able to operate at the broader level. Cross-border governmental cooperation is another matter. Between the Michigan (USA) and Ontario (Canada), for example, there is one of the largest cross-border trade flows in the world, but little institutionalized cooperation since the trade is largely in the hands of multinational corporations who do not look for government intervention, as opposed to the opening of market opportunities. Small and medium-sized business, on the other hand, may look to help in gaining markets and finding partners.

Cross-border cooperation and its success or failure, thus require their own explanation beyond neo-functionalism or economic imperatives. One factor is the existence of political incentives for politicians and officials to open the border. This depends on local political factors. Nationalist entrepreneurs in stateless or cross-border nationalities have an obvious incentive to reach out as part of a strategy of outflanking the nation-state and creating new spaces of identity. This may be part of a strategy of subverting the state for irredentist purposes, but often it represents a realistic view that abolishing the border is not possible but working around it is. Anti-nationalist politicians on the other hand, may regard this with alarm, or accept it as a way of taming nationalism or reducing it to the domains of culture and economic cooperation, reducing its political content. Often the key factor here is not the actual measures undertaken but the political context in which they are interpreted, as dangerous irredentism, or benevolent cultural cooperation. Boosterist local politicians often launch cross-border initiatives hoping for development opportunities but, as noted above, they have to face their own electors and usually those who lose out to competition are more vocal than those who gain. Politicians tend to have short attention spans and, after the initial publicity of reaching out over the border, lose interest unless the cooperation is institutionalized and involves other actors.

Institutional mechanisms and incentives are important, hence the proliferation of cross-border initiatives in Europe under the aegis of the EU and its structural programmes. These provide an incentive, tying cooperation into the broader project of Europe in a way that does not threaten states and giving financial assistance to projects. EU programmes also provide an institutional mechanism while the Council of Europe's Madrid convention gives an outline legal instrument. Within this broad envelope, however, a variety of actors must find their own accommodation, programmes of action and resources.

Another key question is the compatibility of the institutions on either side of the border. Where programmes are undertaken by local and regional governments, they often have quite different powers and resources, or one level might be missing altogether. Budgeting requirements and cycles may differ, making it difficult to raise matching funding, and central governments may intervene more or less capriciously at various stages, with legal, administrative or financial restrictions. More broadly, the tasks of government on either side of the border need to be commensurate if common tasks are to be addressed. Cooperation in Europe seems to be easier, despite the cultural and linguistic differences, than in North America, because of the common conception of government and the welfare state.

Cross-border cooperation operates at multiple levels but three are worth distinguishing, the state, the regional and the local level. Most programmes are organized at the regional and local level, although state elites sometimes try to control them by designating the whole border as a zone of cooperation. In Canada, it is often difficult to distinguish the levels, since the bulk of the national population lives near the border. The local and regional level, however, are favoured by the European Commission whose cooperation programmes are part of its regional policy. The aim is to foster the single market by removing border obstacles but also to exploit local and regional complementarities, uniting economic regions that have been divided by political lines. Region-building projects, tying together new institutions, economic development projects and sometimes social policies, have proliferated in Europe. They represent a new form of boundary-building, not in the hard and rigid way of the old nation-state but nevertheless bringing together economic, social, cultural and political dimensions with the idea that they should be mutually reinforcing. In many ways this is not only compatible with the European project but actively encouraged by the European Commission, which has adopted the new thinking about regional development and the importance of institutional factors, networking and social capital.

Yet the broader logic of economic integration is often sectoral rather than territorial, so that parts of a local or regional economy might be drawn into transnational networks and detached from the rest. Indeed EU Structural Fund programmes, usually assumed to be a major factor in the regionalization of Europe, are often a set of sectoral interventions at multiple spatial levels. When EU cross-border programmes are aimed at the regional level, they may enhance regional capacity and be a resource for region-builders, but again their very cross-border logic may go counter to the region-building project within each of the partners. Economic, cultural, social and political systems may then be pulled apart as they are partially transnationalized. In particular, while functional systems may transnationalize, voting remains a national phenomenon, whether at state, regional or local levels. Interventions at the very local scale may equally sustain region-building, or else encourage a localist pattern of behavior that undermines the larger regional project. Below the grand rhetoric of integration, therefore, we find local politics and administration, trying to match very specific and often small-scale projects to funding, drawing together state, local, public and private as well as specifically cross-border funding. The micro-politics involved here requires minute study but can be a useful corrective to the grander visions about opening of borders. Equally in need of study are the networks of actors in the private sector, social movements and agencies, and the way they are repositioning themselves in relation to the new structure of opportunities. Again, the complementary dynamics of cooperation and competition are the key to understanding.

Cross-border cooperation thus presents as many conflicts and contradictions as it does clear lessons. Unfortunately, much of these were obscured for years by a simplistic form of neo-functional analysis, encouraged by the efforts of the European Commission itself to talk up the idea and to encourage this sort of research. The chapters of this collection thus provide a welcome corrective and a step forward in serious research on a complex subject.

List of Contributors

Dr. Harlan Koff is Assistant Professor of Political Science and the Director of the *laboratoire de sciences politiques* at the University of Luxembourg. He also co-coordinates the University's European Governance program. Previously, he was a Fulbright Scholar at Clersé, Université de Lille 1 and Jean Monnet Fellow at the Robert Schuman Centre of the European University Institute. His research focuses on comparative immigration politics, comparative border politics and comparative regional integration.

Dr. James Wesley Scott is Assistant Professor of Geography at the Free University of Berlin (FUB) and Research Fellow at the Institute for Regional Development and Structural Planning in Erkner (by Berlin). Prof. Scott obtained his Habilitation (2006), PhD (1990) and MA (1986) at the FUB and his B.Sc. at the UC Berkeley (1979). Among his research interests are: urban and regional development policy, geopolitics, border studies, transboundary regionalism in Europe and North America Changes and the spatial implications of Eastern and Central European transformation processes. Recently, he has coordinated European research projects on crossborder cooperation within the EU's Fifth and Sixth Framework Programmes.

Dr. Anastassia Obydenkova (PhD, European University Institute (EUI), Florence) is Senior Researcher in Modern Russian Politics and EU-Russia politics at the Independent Institute of Social and Nationality Problems of the Russian Federation. She is currently involved in three ongoing projects: "The Division of Powers between regional and local authorities in the EU Member-States and Candidates"; "Europeanisation and Democratisation at the Eastern Border of the EU"; "The Quality of Democracy in Central and Eastern Europe". She was a post-graduate researcher at Yale University. Her most recent publications are dedicated to the problems of post-Soviet states (PSSs), EU-PSSs, EU-Russia politics, democratization, and regionalization.

Dr. Carmen Maganda is Coordinator of the binational Border Water Project for the Center for US-Mexican Studies at the University of California, San Diego. She is also a research collaborator in the "Border Politics in Europe and the Americas" project being conducted at the University of Luxembourg. Previously, she was a HERMES Fellow at Clersé, Université de Lille 1. Her research focuses on water management in cross-border basins in Europe and the Americas, social participation in water politics, and elite behavior and decision-making processes in water management.

Dr. Zoe Bray is of British and French nationality. She holds an MA from the University of Edinburgh in Social Anthropology and Sociology, and a PhD in Social and Political Sciences from the European University Institute, Florence. Her research covers the themes of identity, nationalism, borders, minorities and European integration, with a special focus on the Basque Country.

Dr. Monika De Frantz is Research Fellow at London School of Economics and has held research and teaching positions at EUI Florence, University of Vienna, and Bauhaus-University Weimar. With a Ph.D. from European University Institute (EUI Florence), her work focuses on territorial politics, European integration, multi-level governance, urban and regional politics, cultural policy, borders, national minorities, collective action and institutions, combining theoretical and comparative empirical research.

Dr. Daniel Sabet is currently a Visiting Professor at Georgetown University. His research focuses on the role of citizens and civil society in the provision and production of public goods and services. Sabet has lived and worked along the US-Mexico border, holding past affiliations with Mexico's Colegio de la Frontera Norte and the University of California San Diego. In addition to his academic work, Sabet has spent the last two years with the Culture of Lawfulness Project collaborating with police departments in Northern Mexico on the development of rule of law education programs. Sabet received his Ph.D. in political science from Indiana University.

Dr. Michael Keating is Professor of Regional Studies at the European University Institute, Florence; and Professor of Scottish Politics at the University of Aberdeen.

Regionalism & Federalism

The contemporary nation-state is undergoing a series of transformations which question its traditional role as a container of social, political and economic systems. New spaces are emerging with the rise of regional production systems, movements for territorial autonomy and the rediscovery of old and the invention of new identities. States have responded by restructuring their systems of territorial government, often setting up an intermediate or regional level. There is no single model, but a range, from administrative deconcentration to federalization. Some states have regionalized in a uniform manner, while others have adopted asymmetrical solutions. In many cases, regions have gone beyond the nation-state, seeking to become actors in broader continental and transnational systems.

The series covers the gamut of issues involved in this territorial restructuring, including the rise of regional production systems, political regionalism, questions of identity, and constitutional change. It will include the emergence of new systems of territorial regulation and collective action within civil society as well as the state. There is no *a priori* definition of what constitutes a region, since these span a range of spatial scales, from metropolitan regions to large federated states, and from administrative units to cultural regions and stateless nations. Disciplines covered include history, sociology, social and political geography, political science and law. Interdisciplinary approaches are particularly welcome. In addition to empirical and comparative studies, books focus on the theory of regionalism and federalism, including normative questions about democracy and accountability in complex systems of government.

Series Titles

P.I.E. Peter Lang – The website

Discover the general website of the Peter Lang publishing group:

www.peterlang.com